T0074626

IN A FLIGHT OF STARLINGS

IN A FLIGHT OF STARLINGS

The Wonders of Complex Systems

GIORGIO PARISI

In collaboration with Anna Parisi

Translated by Simon Carnell

PENGUIN PRESS

NEW YORK

2023

PENGUIN PRESS
An imprint of Penguin Random House LLC
penguinrandomhouse.com

Originally published in Italian in 2021 as *In un volo di storni* by Rizzoli, Milan

All images by Studio Editoriale Littera, Rescaldina (MI)

ISBN 9780593493151 (hardcover)
ISBN 9780593493168 (ebook)

Printed in the United States of America
1st Printing

DESIGNED BY MEIGHAN CAVANAUGH

To my wife, Daniella Ambrosino,

who has always been by my side

CONTENTS

PREFACE

At this moment in time, perhaps more than any before it, it is essential that the public have a fundamental understanding of the practice of science—that is to say, not only the results at which scientists arrive but *how* they do so. Examples of this urgency surround us. To name what is probably the most urgent, we must make some essential decisions in order to fight climate change. For decades, science has been warning us that human behavior is setting the stage for a dramatic rise in the temperature of our planet. But science alone is not enough. Forewarned is forearmed, as the saying goes—but only if we heed the warning and act on it.

Unfortunately, the actions taken by governments to date have not been up to this challenge, and the results so far have been extremely modest. Now that climate change is starting to affect people's lives, there is perhaps a stronger reaction, but we need much more forceful measures to be taken. Political decisions are needed, especially by the wealthiest countries. We need to transcend our myopic national interests in order to solve global problems. To name a second example, COVID has taught us that we are all connected: what happens in game markets or in the Amazon rain forest deeply affects us all. The recent pandemic has also shown that it is not easy to respond effectively in time. We have seen how measures to contain the pandemic were often taken too late, only when they could no longer be postponed. I remember the head of one European government saying that "we cannot go into lockdown until the hospitals are full; otherwise people will not understand the decision to do so."

Our generation is on a road fraught with dangers. It is as if we were driving at night: the sciences are our headlights, but it is the responsibility of the driver to not leave the road and to take into account that the headlights have a limited range. In order to use those lights in the first place, however, we need to have trust in science.

We have seen during the pandemic the tragedy of the many people who have died refusing to be vaccinated, despite the millions of COVID-related deaths. This has happened thanks to a rejection of science that becomes even more serious when it occurs in relation to climate change.

If citizens and politicians do not trust science, we will move inexorably in the wrong direction, and the struggle against any number of global ills—global warming, infectious disease, hunger and poverty, the depletion of the planet's natural resources—will fail.

How can we promote this trust? Clearly it is not enough for scientists themselves to simply say "trust us." It is also not enough to write scholarly articles about how science works. We must, as the saying goes, show our work: demonstrate in an engaging way how scientists toil, doubt, succeed, and fail. It is important to understand how scientific consensus is achieved, how individual discoveries become validated by the scientific community.

To this end, in this book I have told something of my own story through select reports on significant episodes in my scientific life. I begin with my research into the flight of starlings, those remarkably beautiful murmurations (their flocking behavior) that the physics of this century is only now in a position to explain. I wanted to start there to emphasize how difficult it is to understand the many phenomena that we observe almost daily and to convey that complexity is not about what happens in laboratories. It is about what happens all around us. Our job as scientists is to illuminate for everyone the truths that we discover.

One

IN A FLIGHT OF STARLINGS

The question of interaction is a crucial one in many areas, including for the purposes of understanding certain psychological, social, and economic phenomena. The work described in this essay focuses on how each member of a flock of birds is able to communicate in order to move in a coherent way, producing a single entity that is at once collective and multiform.

It is fascinating to observe the collective behavior of animals, be it that of flocks of birds, shoals of fish, or herds of mammals. At sunset we can see starlings forming fantastic patterns, thousands of milling black specks against the changing colors of the sky. We witness them all moving in unison, without touching or dispersing—avoiding obstacles, moving apart and coming together again, continuously reconfiguring their spatial arrangement, almost as if an orchestra conductor were giving order to everything that is happening. We could watch these birds endlessly, because the show they put on is perpetually altered and renewed, taking different and unexpected forms.

When confronted with all this beauty, the professional curiosity of a scientist cannot but be piqued; wheels begin to turn and questions form. Is there really something like a conductor at work behind this collective behavior, or is it self-organized? How can information be transmitted so rapidly throughout the flock? How is it possible that the shapes made by the collective can change so quickly? How are the different speeds and acceleration rates of the birds distributed within the murmuration? How can they maneuver together without bumping into each other? Might there be simple rules of interaction between starlings that account for such complex and variable collective movements as the ones we see in the skies?

When we are curious and want to know the answers to certain questions, we begin by looking for them in existing research: previously in books, nowadays online. If we are lucky, we find the answers. But when there is no answer to be found because no one has fathomed it yet, and if we are *truly* curious, we begin to wonder if we might be the ones to uncover it. The fact that no one else has succeeded is scarcely off-putting. This, after all, is the scientist's calling: to imagine or to do what no one has done before. But we cannot spend our lives attempting to open locked doors that no one has a key to. Before beginning, it is crucial to understand whether or not we have the competence, technical skills, and tools that will allow us to accomplish the job. Even then, nobody has any guarantee of success. We have to throw ourselves headlong into the task and aim high. Still, if the target is so high that we have no chance of reaching it, then it's better not to begin.

THE FLOCKING OF STARLINGS held a particular fascination for me, because it was connected by a common thread not only to my research but to so many other studies in modern physics that attempt to understand the behavior of systems composed of a large number of interacting components or actors. In physics, depending on the context, the actors can be electrons, atoms, spins, or molecules. These actors may have very simple rules of behavior that when taken together give rise to much more complex collective behavior. Beginning in the nineteenth century, statistical physics attempted to respond to questions related to and arising from such interaction. Questions such as why a liquid boils or freezes at certain temperatures, or why certain substances (such as metals) are good conductors of electric current and heat, whereas others are insulators. The answers to some of these questions were found over time. We continue to look for the answers to others.

In all such physics problems, we try to understand quantitatively how collective behavior emerges from simple rules governing the interaction between individual actors. The challenge here is to extend the applicability of statistical mechanics techniques from inanimate entities to living animals such as our starlings. The results would be of interest not only to ethology and evolutionary biology but might also, in the longer run, lead to a greater understanding of economic and social phenomena. In these cases too we are dealing with great numbers of individuals who influence each other. In general we often need to understand the connection between the behavior of single individuals and collective behavior.

The great American physicist Philip Warren Anderson, a Nobel Prize winner in 1977, expounded this idea in his provocative 1972 article "More Is Different," maintaining that an increase in the number of components in a system creates a change that is not just quantitative but qualitative. The main conceptual problem facing physics was to understand the relationship between microscopic rules and macroscopic behavior, in a great variety of different systems. Anderson played a key role in the development of complexity theory, and we will come across him again in a later chapter.

IN ORDER TO EXPLAIN SOMETHING, we must first know about it, and in this case we were lacking key data. We needed to understand the spatial movements of flocks, but this information was unavailable at the time. In fact, the enormous quantity of video footage and photographs that *was* available (and that can now be easily found on the internet) was all taken from a single point of view or camera angle, and was therefore devoid of three-dimensional information. In this respect we were like the prisoners in Plato's cave, who could see only the two-dimensional shadows of things projected on the walls of their prison, but not the three-dimensional aspects of the objects that made them.

It was precisely this difficulty that made the study of the

movement of starlings so interesting for me: it was a complete project in itself. It involved designing the experiment, collecting and analyzing data, developing computer modeling for simulations, and interpreting the experimental results.

We knew that the methods of statistical physics, which have always been my field of research, would be indispensable for the three-dimensional reconstruction of the movements of the birds, but the thing that really attracted me was the planning and realization of the experimental part of the project. We theoretical physicists usually work at a distance from the laboratory, with abstract concepts. Solving a real problem involves keeping a large number of variables under control—ranging, in this case, from the resolution of the focal lengths of camera lenses to the optimal positioning of the cameras, and from data storage capacity to data analysis techniques. Each element can affect the success or failure of the experiment, and when you plan at your desk (or on the training ground), you have no idea how many problems will occur in the field, or out on the pitch. Which is why I never liked being too far from the lab.

Starlings are extremely interesting birds. Centuries ago, they spent the warmer months of the year in northern Europe and wintered in North Africa. Today winter temperatures are higher due to climate change, and our cities have also become much warmer—partly on account of their growth in size, partly due to the increase in sources of heat, such as central heating and traffic. Many starlings no longer cross the Mediterranean, overwintering instead in various coastal Italian cities, including Rome, where winters are much milder than they used to be.

The starlings arrive at the beginning of November and leave at the beginning of March. They are usually quite punctual in these relocations, and their migration probably depends not so much on temperature as on astronomical factors such as the number of daylight hours. In Rome they find evergreen trees for the night, to protect themselves from the wind. During the day, food is scarce in the city, and they go in flocks of about a hundred to find it in the countryside beyond the Grande Raccordo Anulare, the main motorway that circles Rome. Starlings are social birds accustomed to living in groups: when they alight in a field, half of them feed calmly while the other half keep watch for predators. When they move on to the next field, the roles are reversed. In the evening they return to the warmth of the city, and before settling in the trees, they form into flocks of huge numbers of individuals whirling in the sky above the capital. Ultimately, they are still birds that are sensitive to the winter air. After nights when a cold north wind has blown, it is common to find many starlings dead beneath trees that have not provided them with sufficient protection.

So a good choice of roosting site is literally a matter of life and death. It is highly probable that their celebrated aerial choreography represents a signal, visible from a great distance, indicating the location of a suitable place to spend the night. It is like waving a huge, extremely conspicuous signaling flag: I myself have seen, with the naked eye, the evolutions of starlings about ten kilometers away—grayish specks that moved almost amoeba-like against a sky that still had a thin strip of white light just above the horizon. The first small groups to arrive from the countryside begin to dance in a way that becomes ever more frenetic as the light fades. Gradually the latecomers join, until in the end flocks of thousands of individuals are formed. Half an hour after sunset, when the light has disappeared, they suddenly throw themselves down onto the chosen trees.

A peregrine falcon often appears, looking for its supper. If you're not on the lookout for it, it can easily go unnoticed: your attention is focused on the starlings. Despite the fact that the peregrine is a raptor with a wingspan of one meter and can reach dive speeds of more than two hundred miles per hour, the starlings are not easy prey. An airborne collision with a starling can easily fracture a peregrine's wing, resulting in certain death. The falcon cannot enter the flock, but seeks instead to pick off individuals at its margins. The starlings react to the threat of the falcon by getting closer to each other, closing ranks and rapidly changing direction in order to elude its lethal grasp. Some of their more spectacular shape-shifting is caused precisely by evasion of repeated attacks by peregrine falcons that need to make numerous attempts before catching their prey. It is probable that many of the behaviors of starlings are due to the necessity of coping with these fearsome predators.

TURNING TO OUR PROJECT, the first difficulty was in constructing a three-dimensional image of the murmuration and its shape by combining various simultaneous photos. In theory this should have been easy, and the problem resolved in a simple way: everyone knows that to see in 3D, you just have to use two eyes. Looking simultaneously from two points of view, even if they are very close to each other, like our eyes, allows the brain to "calculate" the distance to an object and therefore to construct three-dimensional images. With just one eye, we lose our sense of depth of field. You can easily test this by closing one eye and attempting to pick up with one hand an object that is in front of you: your hand will seek it further away or closer than it actually is. If you try to play tennis or table tennis with one eye covered, you are certain to lose. Our experimental system only works, however, if we are able to match the bird in the left-hand camera with the same bird in the right-hand one—an operation that becomes a nightmare when there are thousands of birds in each photo.

We clearly had our work cut out for us. In the scientific literature of the time, three-dimensional photos had been made of no more than twenty animals at once, by manually identifying their left-right correspondence. We needed to construct several thousand photos, and each one of these photos was of thousands of birds. Obviously it was not possible to do this manually, so we needed to delegate this work to a computer.

Tackling a problem without adequate preparation is a recipe for disaster. Around 1990, at the suggestion of a young physics researcher, Renata Sarno, we had begun to reflect on the problem, but managing hundreds of thousands of analog photos on film was too complicated. Fifteen years later, with the advent of high-quality digital photography, the situation had changed and we took up the project again. We set up a group in which there were not just physicists—my mentor Nicola Cabibbo; two of my best students, Andrea Cavagna and Irene Giardina; and me—but also the ornithologists Enrico Alleva and Claudio Carere. Together with an economist, the late Marcello De Cecco, in 2004 we made an application to the European Union for funding. The application was successful. We could begin, involve undergraduates and doctoral students in the project, and buy equipment.

We placed our cameras on the roof of the Palazzo Massimo, the home of the beautiful Museo Nazionale Romano, overlooking the large square in front of Rome's main train station, which at the time—the first data was collected between December 2005 and February 2006—was one of the starlings' favorite roosts. We used high-end commercial cameras because video cameras were still too low in definition. Two cameras situated twenty-five meters apart allowed us to determine the positions of two starlings several hundred meters away, with an accuracy to within ten centimeters: precision enough to distinguish starlings flying about a meter apart from each other. We added a third camera a few meters from one of the others, which helped us when two birds overlapped in one of those cameras. This third lens provided fundamental information in cases where reconstruction was particularly difficult.

The cameras clicked simultaneously, within a millisecond of each other (we had built a simple electronic mechanism to control them), taking five frames per second. In actual fact, at every station there were two cameras in contact with each other, in order to double the frequency of images, giving us ten images per second. In the end this was not so inferior to a video camera, which normally takes twenty-five to thirty images per second. We were using still cameras, but in reality we were obtaining short films.

I'll leave aside here all the technical problems associated with the alignment of the cameras (accomplished using a taut thread of fishing line), the calibration of focal length, and the rapid storage of a huge number of megabytes of information. In the end we succeeded, not least thanks to the tenacity of Andrea Cavagna, to whom I had entrusted management of these operations as both a much better organizer than I was and someone less distracted by numerous other tasks.

We needed not only to make films in 3D, a technical feat difficult enough in itself, but to then reconstruct the three-dimensional positions. With 3D movies in cinemas, this is easily done: every eye sees what has been filmed by a camera, and then our brain, evolutionarily selected for over millions of years, is fully capable of arriving at three-dimensional vision by locating in space the objects that it sees. We were faced with a similar task using computer algorithms instead—the second part of our challenge. We deepened our repertoire of statistical analysis, of probability, of sophisticated mathematical algorithms. For months on end we feared that we would not make it: sometimes when tackling a problem that is too difficult, we find our hands tied in unforeseen ways. Fortunately, through hard work, we managed to devise the mathematical tools to resolve the difficulties one after another, and a year after the first quality photo was taken, we had the first three-dimensionally reconstructed image.

. . .

EVEN IF THE BEHAVIOR of starlings is a subject for biologists, the quantitative study of the three-dimensional movements of individuals requires the kinds of analyses that can be accomplished only by physicists. The simultaneous analysis of thousands of birds in hundreds of photos in order to reconstruct the trajectory of individuals in space and time is typical of the kind of thing we do. The techniques required for such analysis have much in common with those that have been developed to solve statistical physics problems or to analyze massive amounts of experimental data.

After almost two years of work, we were the only people in the world to have many three-dimensional images of groups of starlings. By simply observing them we have learned a lot. When we look with the naked eye at starlings from the ground, one of their most striking characteristics is the speed with which the flock changes shape, in a way that is difficult to describe to anyone who has never actually seen it happen. In the sky, moving objects of variegated shapes suddenly become smaller, narrower, then get bigger again, change, become faint—almost invisible—and then darker. There is enormous variation in both their shape and density.

Many simulations of their flight, attempts to imitate this behavior using computers, began with flocks that were basically spherical. The first three-dimensional photos, however, revealed that flocks were in fact more disklike. It is precisely for this reason that they change shape so rapidly: a disk-shaped object can appear very large and round if viewed flat, and decidedly thinner when viewed from the side. The enormous and extremely rapid changes in shape and density are therefore an effect of changes in orientation with regard to us—an explanation that had been proposed by Nicola Cabibbo before we did the experiment, though without the observational data, we were not yet able to prove that this was correct.

We were extremely surprised to discover that the density at the edges of the flock was nearly 30 percent greater than at the center. Starlings are closer together at the margins compared with the middle: a bit like what happens on crowded buses, where frequently the crush is greatest near the doors where passengers who have just got on accumulate, together with those who are about to get off and others still who want to continue their journey. If we supposed naively that the birds in a flock were like attracting particles, we would expect the density to be greatest at the center and to diminish toward the edges. In fact the opposite turned out to be the case. The murmurations also have very defined boundaries: it is rare for a single bird to detach from the group. It is probable that this behavior has a biological origin in reaction to predation by peregrine falcons. An isolated bird is easy prey, and the closer together the birds are at the edges of the flock, the more difficult it is for the falcon to pick one of them off. The birds at the margins tend to bunch up as a defense mechanism, whereas those in the center do not need to huddle in order to feel safe: they are already protected by their fellows at the edges.

Looking at the first photos, we observed that each individual had more distance between itself and the birds directly in front and behind than it did between itself and the birds on either side. It is rather like what happens with cars on a motorway: it is perfectly normal for two cars to be no more than a couple of meters apart laterally, even at very high speeds, whereas it is extremely inadvisable to get that close to the car immediately in front.

This tendency of the birds to distance themselves from those in front and to stay closer to those on either side is present both in more compact flocks (with an average distance of around eighty centimeters) and in flocks that are much more spread out (with an average distance of around two meters). This phenomenon does not depend on the distances between birds. It is reasonable to assume that it is not due to a problem of dynamics, as when two planes must keep apart in order to avoid each other's turbulence; otherwise the effect would be much less when the birds are more distant. It is due rather to the way in which they orient and mirror each other while avoiding collisions.

THIS CHARACTERISTIC OF the positioning of starlings has allowed us to reach a truly unexpected conclusion: the interaction between starlings depends not so much on the general distance between them as on the connections between the closest birds. This seems only natural: if I go for a run with friends and we turn right, in order to keep pace while doing so, my attention is fixed on the runner nearest to me (one or two meters away), and it hardly matters what friends further away than this are doing.

With hindsight this seems obvious, yet it is striking how in both physics and mathematics there is a lack of proportion between the effort needed to understand something for the first time and the simplicity and naturalness of the solution once all the required stages have been completed. In the sciences as in poetry, there is hardly a trace in the finished product of the arduous work that the creative process has demanded, or of the doubts and hesitations that have been overcome in order to achieve it.

Physics, from Newton's law of universal gravity onward ("the force of gravity between two bodies is inversely proportional to the square of the distance between them"), is wholly familiar with interactions that depend on distance. It hardly occurred to us that distance might have a marginal role to play in determining the strength of interaction—until the experimental data hit us in the face.

How did we get there? We first expressed quantitatively the previous observation that birds tend to respect a greater "safe distance" between themselves and those in front than with those to the side: in this way we defined a quantity that we have called *anisotropy* (in physics a quantity is anisotropic if it has different values in different spatial directions). When pairs of birds were close together, the anisotropic value was high, whereas for distant birds the value was close to zero. Up to this point we were happy: we expected that distant birds would not have information on their respective positions, and it was only logical that there would be no difference between their lateral and frontal linear distances.

But we ran into serious problems when we compared the anisotropy between birds at the same distance from each other in different sequences of photos. Nothing made sense: sometimes the anisotropy for birds at a distance of two meters was very large, whereas in other groups of photos the anisotropy at the same distance was negligible. In the end we realized that comparing the behavior of two birds at the same distance in different flocks doesn't work, because the distance between the closest birds varies greatly from flock to flock.

We changed tack: for every bird, we identified its first neighbor, then its second, and then its third, and so on. We found that anisotropy was high among first neighbors, somewhat lower among second neighbors—and virtually nil between the first and seventh birds. At first glance it would seem that there is no more information here than before: anisotropy diminishes with distance. Things change, however, when we compare flocks. Anisotropy was the same for pairs of first neighbors in different flocks, even though the average distance between these pairs varied, to the extent that in some flocks it was more than twice the first-neighbor distance in others. At this point no great intellectual effort was required: the data forced us to see an interaction between birds that did not depend on the absolute distance between pairs but on the relative ratios of the distances. In technical terms, we saw that their behavior depended on topological rather than metric distance.

This was the result of the first stage of our work, in 2008. Since then a lot of water has passed under the bridges of the Tiber. The composition of the research group has changed; I have started working full time on spin glasses; new funding has been provided, and we have bought new equipment that is much more advanced. There are cameras on the market now that are capable of taking 160 frames per second, at a resolution of four megapixels.

We now have new ideas and new algorithms, and we are able to determine, within several hundredths of a second, the precise moment at which each individual bird begins to turn when the flock as a whole turns. Almost always there is a small group at one side of the flock that initiates the turn, and in a very short space of time—a few tenths of a second for small flocks, a whole second for large ones—all the other birds follow suit. After analyzing the data and drawing on complex theory, we understood in great detail the quantitative behavior of a flock, even during a turn. The birds follow simple rules that we discovered from the measurements and reconstructions we made, and their movements are prompted by the positions of their neighbors. Information about turns passes very quickly from one bird to another, as if by incredibly rapid word of mouth.

Our research has completely changed the paradigm used when studying flocks and herds. Before our work, it was taken for granted that interaction depended in some general way on distance. We found that birds were equally affected by other birds that were one meter or two meters away. But perhaps the most interesting outcome was the concrete proof that it was possible to track the positions of thousands of birds at the same time, and to extract from this knowledge information that would have a useful bearing on the wider understanding of animal behavior.

Our results were possible because we used quantitative techniques for the statistical study of the behavior of a few thousand animals. We defined new standards of investigation in biology by using techniques originated and developed in statistical physics to solve complex and disordered problems. Not all biologists appreciated this incursion into their territory: some have shown themselves to be very interested in the results, while others have found our investigations to be too short on biology and top-heavy with math. The work was rejected by various journals that are probably kicking themselves now. After the great success of our first article, which was cited in almost two thousand scientific publications, many others have followed.

Biology is going through a period of great change: the accumulation of huge amounts of data makes the use of quantitative methods not just useful but essential. These methods can be used both appropriately and inappropriately, depending very much on the context. In particular, in ethology, the study of animal behavior, too much mathematics can easily produce a negative reaction. Ethologists look for the reasons behind certain behaviors, whereas quantitative methods are purely descriptive and therefore do not reach the heart of ethological research.

The spirit of many scientific disciplines has changed over the years, but this has happened only through passionate debates over which methodologies are scientific and relevant and which should be rejected as incapable of answering the real questions posed by the discipline. The skeptical words of the great founder of quantum mechanics, Max Planck, come to mind on this subject: "A new scientific truth does not triumph by convincing its opponents and making them see the light, but rather because its opponents eventually die, and a new generation grows up that is familiar with it." I am more optimistic than Planck: I believe that with a lot of good faith and a lot of patience, at least in the majority of cases, it is possible to arrive at shared conclusions. Or at the very least to clarify where our disagreement comes from.

Two

PHYSICS IN ROME, AROUND FIFTY YEARS AGO

It is important to preserve the memory of the past, especially in the field of science, and that is why I would like to tell you about my first years at the university and what physics was like at the time. What follows is not a history; it is limited to my own personal memories: the recollections of a theoretical physicist who is interested in elementary particles.

At eighteen, in November 1966, I began my first year of undergraduate physics. Back then, first- and second-year students were not free to roam about the physics department at the Sapienza University in Rome, where I had matriculated. We attended lectures on general and experimental physics, but we were required to enter and exit the building through the back door. It was considered unseemly to have crowds of students going in and out of the main entrance at the front, and that entrance was patrolled with an iron hand by Agostino, the veteran, beady-eyed doorman (or better still, goalkeeper) on whom nothing or no one seemed to be lost. Agostino blocked first- and second-years by demanding what business they had there. Because most of the time, except on special occasions, those students did not have anything particular to do, and they were firmly directed around the back.

There were about four hundred of us attending the first-year lectures, and there were no microphones. The professors had to shout to be heard. The lectures on general physics, by far the most important and formative course, were given in alternate years by Edoardo Amaldi (one of Enrico Fermi's colleagues in Italy) and Giorgio Salvini (an experimental physicist who directed the construction of the first large electron accelerator in Europe, near Rome). I was taught by Salvini, who, compared with the reserved and formal Amaldi, was something of a showman. Salvini once sat in a swivel chair and proceeded to rapidly spin while holding two heavy iron dumbbells, one in each hand, to demonstrate how he would spin faster when he closed his arms and slower when he opened them again. It's a phenomenon that ballet dancers are perfectly familiar with: to execute a pirouette, they must begin with arms open and close them when turning. The lesson ended with a statement about the law of conservation of angular momentum, which explained what we had just witnessed in the form of a spinning professor.

You went through the main door above all to go to the Fisichetta (Little Physics) lecture hall, so called to distinguish it from the general physics lab, called the Fisicona (Big Physics). Exercises were carried out in a labyrinth of subterranean rooms—which I remember, with their concrete floors, being dank—and in every room there was a different experiment to be done: atmospheric pressure, the fall of a body on an inclined plane with little friction, the measurement of the energy expended in order to melt ice. We went there in groups of thirty: there were ten desks to a room and three of us to a desk, in a partnership that would last for the entire academic year. With this setup it was difficult to meet older students. We had no contact whatsoever with anyone who was not in our own year.

EVERYTHING CHANGED IN 1968. Not just the university, but politics in general in Italy, as well as across Europe and throughout the world. An enormous radicalization of society, from top to bottom, was reflected in changes in customs of every kind. People like me who came from moderately conservative backgrounds and who voted for the Liberal Party or the Christian Democrats found themselves in a maelstrom of social conflict, veering toward Marxist ideas. Rivers of ink have been spilled on the history of '68, trying to explain its causes and consequences, and I won't add to that here. I want to describe instead the effects of '68 within the Physics Institute. For me, everything began with a meeting in the packed main lecture hall, with more than six hundred of us crammed into a place with just three hundred seats. The meeting went on all afternoon and continued well into the evening, and at nine p.m. we voted on whether or not to occupy the buildings. The occupation was voted for with a huge majority, by around two to one. We the students had voted for it, and what happened in the Physics Institute was now our responsibility—as well as the responsibility of those who by voting "no" had conferred legitimacy on the process.

When Giulio Caradonna, one of the leaders of the neofascist party, descended on the university accompanied by squads of supporters wielding long hard sticks and wrapped in the Italian flag, the director of the department, Giorgio Careri, who was completely overwhelmed by what was happening, was extremely concerned for the safety of the library on the second floor—not least because the fire extinguishers had been requisitioned by us and brought to the lecture hall for improvised use in repelling the assailants. Approaching the students who were on duty at the entrance of the Institute (and clearly of lesser concern), he remarked: "If the inevitable happens, make sure to let it happen on the first floor."

With the occupations and sit-ins, all barriers between students of different years had fallen—as well as those between students, teaching assistants, and young lecturers. There was a real sense of camaraderie among us. (One of the lecturers, the nuclear physicist Paolo Camiz, was a singer-songwriter who performed at Folkstudio—a small place in central Rome where a little-known Bob Dylan had played in 1963.) There were two reading rooms. In one, surrounded by a floor-to-ceiling collection of journals going back decades, respectful silence reigned. The other was much noisier: we talked, laughed, even played bridge together there in the late afternoon (the traditional Italian card game scopa was not considered serious enough for any self-respecting physicist). The department was used much more than it is now: in the evening, after nine p.m., a back door opened and the working students who could not attend during the day were let in.

From my point of view, this was a world infinitely younger than the physics departments of today. Obviously I was myself much younger—more than fifty years younger—and naturally, unlike now, I also kept company with the young. And yet the Physics Institute really was objectively younger back then. At the time, Edoardo Amaldi, the great figurehead of Italian physics who was affectionately referred to as "Father," was just sixty years old. Under Amaldi, the main professorships were held by Giorgio Salvini, Marcello Conversi, Giorgio Careri, and Marcello Cini, who were all around fifty or under, and markedly younger than current professors.

Nicola Cabibbo had arrived at the department in 1966. A full professor at the age of thirty-one, he found glory with his theory of weak interaction, based on the so-called Cabibbo angle, a discovery that might easily have won him a Nobel Prize. It was the spearhead of all Italian theoretical physics. In 1968, at thirty-three years old, he was the same age as Francesco Calogero, who received the 1995 Nobel Peace Prize as secretary-general of the Pugwash group—a nongovernmental organization founded with the aim of ensuring scientific development that was compatible with world peace.

Many of the assistant lecturers and researchers in theoretical physics were even younger, no more than thirty years old. There were certainly exceptions to this rule, such as the theoretical physicist Enrico Persico, who was friends with Enrico Fermi and died in 1969, at sixty-eight. I did not have much to do with them, however, since unlike today most of the important teaching was done by professors who were in their late forties.

This is not just the impression of one young student; there is a good historical explanation for it. In the 1950s there was an explosion of Italian universities, which were in the process of becoming the institutions for the masses that we know today. Physics in particular had developed strongly and received substantial funding, thanks in part to Amaldi, who was the first secretary-general of CERN (Conseil européen pour la recherche nucléaire, or the European Organization for Nuclear Research): research activity was completely international, and the prestige for it in Italy had been gained from its acknowledgment abroad. The old hierarchies that dominated in other institutions and faculties (headed by the infamous "barons") had lost their power in physics, and the best scientists quickly rose to the top in academia. I was elected to a chair at thirty-two. Tenured positions came soon after graduation. When in 1970, at age twenty-two, I started working at the National Laboratories at Frascati, my twenty-five-year-old friends Aurelio Grillo, a theoretical physicist working on high energy (with whom I played poker back then), and Sergio Ferrara, another theoretical physicist who won the Breakthrough Prize, already had tenure. At that age today, if everything has gone well, you might expect to be roughly halfway through a doctorate.

. . .

WE ARE SO USED to exchanging texts easily and at practically no cost, via the internet or by phone, that it is difficult to imagine what scientific communication was like at this time.

International phone calls were incredibly expensive. A call to the United States would cost 1,200 lire (about two U.S. dollars) per minute, at a time when my first monthly salary as a researcher was 125,000 lire. An hour-long conversation at this rate would have exhausted my wages for the entire month. Fax hardly existed yet. In the physics department there was a teleprinter (actually a telegraph terminal) that was exceedingly heavy, cumbersome, and seldom used.

The telephone was employed only on special occasions. One of the most amusing of these concerns the discovery of the psi particle in November 1974. The particle consists of a charm quark and a charm antiquark; its discovery had a significant impact on the physics of elementary particles, so much so that it was dubbed "the November Revolution." The particle was found at almost exactly the same time by two different laboratories in the United States. The news spread rapidly throughout the world. Researchers at the Frascati laboratories realized that they were also in a position to observe it. The parameters of experiments already in progress were immediately modified, and within a single week, psi was also observed by us, to the general joy of the physicists present.

It was an extremely important result; even though it was obtained after the Americans and based on information that came from their experiments, it demonstrated the tremendous skill of the Italians. It was imperative that we write an article for the most important physics journal, *Physical Review Letters*, and have it published in the same issue in which the American articles would appear. The review's deadline for submissions was approaching and there was no time to lose. Immediately after the discovery, the article was written over a single weekend, and to gain time, in what was virtually an unprecedented way of submitting, it was then dictated over the phone. Even the figures and graphs had to be transmitted orally, with the coordinates for the latter spoken for some obliging person on the other side of the Atlantic to plot. The names of the numerous authors naturally had to be dictated as well, and spelled out over the telephone. This led to some peculiar results. It was Giorgio Salvini himself who took charge of the dictation, and because he said "*S* as in Salvini" throughout, his name was turned into an *S*, and he disappeared from the list of authors. At the point where it should have read "G. Salvini, M. Spinetti," there was "G. S. M. Spinetti" instead. Needless to say, an errata was subsequently required.

Extended correspondence, frequently rich in formulas, is a must in scientific collaborations. Letter writing is a particularly inconvenient means of communication in Italy, given that our postal system functions so poorly. Letters sent by air would take fifteen days to reach their destination. Working together at a distance was virtually impossible: it was necessary to be together in more or less the same physical space.

In the spring of 1970, Nicola Cabibbo invited me and my near-contemporary Massimo Testa to listen to him read a handwritten letter from Luciano Maiani, a theoretical physicist who studied with Cabibbo and had gone to work at Harvard for a year. Maiani had written to inform him of results he had obtained with Nobel laureate Sheldon Glashow and Greek French theoretical physicist John Iliopoulos. The letter has stayed with me, not just on account of the extremely important scientific results it contained, but because of its memorable conclusion: "We have thrown out the baby with the bathwater." In fact the letter was informing us that the research program Cabibbo and Maiani had begun some years previously to try to calculate the Cabibbo angle had come to an end. The angle turned out not to be calculable. But in compensation, the letter contained the basis of what would become, using the initials of its coauthors (Glashow-Iliopoulos-Maiani), the GIM mechanism. Explaining how some interactions between particles were permitted or not, the GIM mechanism predicted the necessary existence of neutral weak currents and charm quarks—predictions that were later experimentally verified in 1973 and 1974, respectively.

MOST SIMPLE CALCULATIONS WERE done on paper, with the help at most of a slide rule that was frequently carried around in a pocket. The slide rule is now a museum exhibit. At the time it allowed us to quickly perform multiplications to two or three places, and was swept away by the advent of the portable calculator. I remember my astonishment when in 1973 I saw such a calculator for the first time. It took my entire monthly salary to buy one.

The computers at the end of the sixties were very different from today's laptops and desktops. They did, however, have one thing in common with them. Once, when a good friend of mine who is a few years older than me saw me carrying a packet of punched cards, he wisely warned: "Be careful, computers are malicious." Despite the efforts of generations, the maliciousness of computers is a characteristic that has never been completely expunged—as we can verify when we experience a crash on precisely that rare occasion when we did not back up the file we were working on.

The main computer was a mighty UNIVAC. Located in the basement of a building a few hundred yards from the department, it was accessible only to technicians. Its memory, not counting add-ons with external disks, was about a tenth of a megabyte—in other words, around a millionth of my current mobile phone's. On the second floor there were machines with keyboards—giant typewriters—that punched the cards containing the programming instructions. On each card there were no more than eighty characters. In the middle of the room there was a terminal: a machine into which we inserted packets of cards that had been laboriously "written" on using perforations. The terminal read these cards very rapidly, at a rate of several dozen per second. After a time that could vary from a matter of minutes to several hours, a fast printer would deliver the results on large sheets of paper. Frequently you might hear someone exclaiming: "Shit, I left out a semicolon. I'll need to correct the program and start all over again." There was a queue to feed the cards in. Some people came with small packets of just over a hundred cards; others arrived with thousands carried in a special container, a sort of elongated drawer. A colleague once stumbled and spilled the entire contents of such a container, holding a meter's worth of cards. "There goes the data analysis," he sighed. His task was three-quarters finished—but putting these jumbled programming cards back in order would present an endlessly time-consuming puzzle. He decided to make do with partial data, to end that particular research and move on to other problems instead.

It was scarcely conceivable that data might one day be recorded digitally by computers: there were no machines capable of doing this, and there were no interfaces between instruments of measurement and the mega computer. We went on writing out longhand the data that the instruments recorded. In one particular case, in order to analyze very rapid data, we used one of the most recent advances in technology: a roll of thermal paper that paid out in the form of a tape, at a speed of one meter per second, while a hot pen transcribed the data—exactly like a cardiogram, only much faster.

In particle physics, spark chambers up to several meters across were often used. The passage of a particle inside the chamber causes sparks that make it possible to reconstruct its trajectory. The sparks are photographed, then their coordinates are plotted. This operation (called "scanning") was carried out by assistants, invariably women, who would project the photos onto large tables and manipulate arms like those on a pantograph. When the arm was in the right place, the assistant pressed a button that would punch and print a card. These women worked in a large room on the third floor. Their boring task was fundamental to all experiments in particle physics.

. . .

IN MY CIRCLE AS a young student, the theory of elementary particles was considered the ne plus ultra of physics. Many good friends who were a year older than me were unable to do their theses with Nicola Cabibbo because he was so popular with undergraduates. They had to choose another thesis in another field instead, and despite the fact that the professors who supervised them were also among the best in Italy, it was for them a mere second best and already a kind of failure.

Why did the theoretical physics of elementary particles enjoy such prestige at this time? Enrico Fermi's vital legacy was still very strong in Rome, as were the links with CERN in Geneva, the most important center of particle physics in Europe and perhaps the world. But these two factors alone did not fully account for its appeal. There was an aura of mystery around the theory of particle physics. By now we all know of the existence of quarks: held together by gluons, they are the constituents of protons and neutrons, and there is a theory—quantum chromodynamics (QCD)—that allows us to calculate their properties.

At the time, almost nothing about quarks was understood. From the 1930s onward the proton and neutron were known, and gradually throughout the fifties and sixties it was discovered that there are many other particles, ones that are difficult to observe on account of having such brief half-lives: an endlessly exterminated family of particles (that today we call baryons), the only members of which do not decay extremely rapidly being the proton and neutron, because they are the lightest. The proton and neutron did not appear to have any other particular characteristics.

The fact that there is an entire populous family of similar particles, and that some types of decay were observed and not others, led us to believe that these particles were formed by components that combined in a variety of ways to generate different objects. The almost infinite variety of chemical substances arises from the combination of a hundred different types of atoms. The atoms are made of nuclei and electrons; the nuclei consist of protons and neutrons. But what were the protons and neutrons made of?

It was not a question that was easily answered, and there were no obvious clues. A revolutionary solution was proposed in 1962 by the American Geoffrey Chew: the so-called "bootstrap" theory. It's a word that is often used today in the jargon of programmers to refer to the process of self-starting computers, but was used at the time by just a handful of superspecialized technicians. It derives from the saying that you cannot "lift yourself up by your own bootstraps"—which, if you have not yet tried it, can be easily verified. According to bootstrap theory, every particle is in some way composed of all other particles; there is a "democracy" among elementary particles, and none is more fundamental than any other. The age-old search for the constituent elements of matter (one of the first answers to which was "water, air, fire, and earth") had come to an end; there were no longer constituent elements of matter, only relationships between the various particles. It was an idea that proved to be enormously successful. In *The Tao of Physics*, published in 1975 when the bootstrap philosophy was already on the wane, Fritjof Capra linked it to Eastern philosophy, though to me it had always seemed instead like an echo of Hegelian idealism.

There were so many schools of thought seeking to bring order to the enormous quantity of data that was available, using various principles such as symmetries of all kinds, the impossibility of transmitting information faster than the speed of light, and so on. These were schools that did not talk to each other and that had limited objectives. Bootstrap was the most radical, aiming at producing a complete theory.

At this point an informed reader might be forgiven for wondering why they did not use a theory based on quarks. Quarks had been known about since 1964, when they were first proposed by Nobel laureate Murray Gell-Mann and George Zweig, with Oscar Greenberg adding the concept of "colors" just a few months later (every type of quark exists in three different so-called colors). Quarks were first introduced as mathematical simplifications, and the fact that no one had succeeded in observing them, despite painstaking experimental research, had rendered their existence not very credible. What predominated instead was the "philosophy of the pheasant and the veal," in reference to the metaphor Gell-Mann employed in a famous work of 1964, after a discussion with Valentine Telegdi. Gell-Mann had used the quark model to formulate a series of equations, but for him these equations were much more important than the model itself, which was simply a means of obtaining them. At this point we might forget the quark model and just keep hold of the final equations. The method employed was like the one used in French cuisine whereby a piece of pheasant is cooked between two slices of veal: when ready, only the pheasant is brought to the table, and the veal is thrown away. Even those who took the quark model seriously could use it only to a very limited extent.

Gradually, toward the end of the 1960s, things began to change: new experimental data arrived, the theory was refined, and eventually it was apparent that colored quarks and gluons could explain the experimental data. This view triumphed with the so-called November Revolution of 1974, when the discovery of the psi particle and its strange properties tipped the balance conclusively toward the theory as we know it today.

But what became of the bootstrap?

At one of the most important research centers in the world, the Weizmann Institute in Israel, there was a formidable group of physicists led by the brilliant Argentinian Hector Rubinstein. Under his guidance, Miguel Virasoro, Gabriele Veneziano, Marco Ademollo, and Adam Schwimmer began a series of studies of particle physics from which string theory arose. The fundamental step toward this theory would be taken by Veneziano with the first string model in 1968, but these preliminary studies were crucial in forming the conceptual framework in which that model could be conceived. Stimulated by Veneziano's work, Virasoro extended the theory by introducing the model of closed strings. These results triggered a wave of interest, and gradually it was discovered that formulas could be derived from postulating that matter is constituted by a string (an elastic one) and that the various particles correspond to its oscillations. Unfortunately, though, these strings did not directly describe the observed particles.

In 1974, Joël Scherk and John Schwarz realized that string

theory could be used as a point of departure for describing gravity in a quantum scenario (quantum gravity), even though many details were missing back then, as they still are even now. It is paradoxical that bootstrap theory, which wanted to eliminate the elementary constituents of matter, became the midwife to a new theory in which everything that exists in the universe (matter, light, and gravitational waves) is made up of strings.

Ideas are often like boomerangs: they start out moving in one direction but end up going in another. Interesting and unusual results can turn out to apply to totally unexpected fields.

Today we have a good understanding of the properties of the proton and other particles, but as far as quantum gravity is concerned, we are in a situation reminiscent of fifty years ago. There are various schools of thought: string theory, loop quantum gravity, and so on. Is one of these right, or do we have to wait for a new theory or experiment that will provide us with unexpected results? What form will the final theory take? It is difficult to say. However much we try to predict the future, the only thing certain is that it will surprise us.

Three

JE NE REGRETTE RIEN

I've never been able to figure out if having a Nobel Prize slip from under your nose at twenty-five is something to be mentioned with pride or one of those slightly shameful secrets that is best forgotten. I lean toward the latter view, but because the story of how it happened is amusing, I'll talk about it anyway. Some effort is needed to explain the scientific context.

We need to think back to the end of the 1960s. The experimental picture regarding elementary particles was clear: the proton, the neutron, and the other particles known at the time interacted strongly with each other. In other words, if you make them collide, their trajectories change, and at very high energies the collision produces many other particles. Collisions in which two protons bounce elastically off each other, like billiard balls, are extremely rare and happen when the energy of the impact is very high.

The rarity of these collisions was explained by a theory in which the proton and the neutron are composite particles. During the collision, they literally fall to pieces and therefore cannot bounce off each other whole. How their fundamental constituents behaved, however—those particles from which the protons and neutrons were formed—remained to be understood. There were two possibilities:

- The collisions in which these particles bounced were frequent, even at high energy. They therefore strongly interacted with each other at all energies. In this case the behavior of matter remained always difficult to understand, and there were no simplifications at high energies.
- The collisions in which these particles bounced were *in*frequent, which is to say that the particles interacted weakly at high energies and became almost invisible to each other. At high energies, the behavior of the constituents of protons and neutrons was easy to calculate: their trajectories did not change, as if there was no interaction. One theory of this kind is what today has become known as asymptotically free. (In the jargon of physics, a theory is *free* when the particles do not deviate from their trajectories, and by *asymptotic* we mean "at high energies.")

An asymptotically free theory had the advantage that at high energies some quantities could be calculated in a rather simple way, so there was a huge number of phenomena that were potentially predictable, much to the delight of theoretical physicists. However, given that the universe was in all probability not designed to make the lives of theoretical physicists easy, this argument did not necessarily imply that the said universe could be described by an asymptotically free theory.

I started working with the first hypothesis, liking it better because it described a situation more difficult to understand and provided more of a challenge to get results. For other scientists it had proven to be, as in Aesop's fable, a question of not wanting to take the grapes because they were "too sour." In fact no one had been able to conceive of a theory in which the possible constituents interacted less and less the more the energy was increased; I believe that the few people who *had* considered the problem had concluded that such a theory probably did not exist. In 1955 the brilliant Russian physicist Lev Landau noticed that in all theories the strength of the interaction increased with an increase in energy, except perhaps in electromagnetic ones in which the field itself was charged (what are called Yang–Mills theories)—though accounts of them were difficult, and therefore it was not yet possible to know if it was true or not. From a technical point of view, Landau had discovered the existence of a function (commonly called *beta*) that controlled behavior at high energies: if the beta function was positive, the interaction always remained strong; if the beta function was negative, the interaction was asymptotically free.

In 1968, Richard Feynman proposed that the known particles were composed of point-like constituents with negligible interactions at high energies, which he called partons because they were parts of matter. And yet despite the strength of this proposal, attempts to construct an asymptotically free theory languished.

It was only in 1972 that Sidney Coleman published a work showing that Landau's conclusions were perfectly justified, even when considering more complicated models than those studied by the Russian physicist. All that was left was to study the Yang–Mills theories in order to understand the sign of the beta function: a negative sign would be an unexpected result with profound consequences for physics. With an irony of sorts, we discovered many years later that this work had already been done in 1969 by the Russian physicist Iosif Khriplovich—and published in a Russian review, a translation of which was in our library. Its unfortunate author was ahead of his time: despite the great elegance and clarity of his account, no one had paid any attention to the result. I discovered it only by chance while looking up another article in the same review.

The importance of calculating the sign of the beta function in the Yang–Mills theories was obvious to me, but I was preoccupied with a different problem (phase transitions) and did not devote much time to it. I remember that after reading Coleman's work in the spring of 1972 I began to think about the sign of the beta function in this theory. One day at my parents' house I was concentrating on the problem while soaking in the bath and staring at the orange marble tiles that covered the bathroom walls, when it came to me all at once: the beta function must be the sum of three distinct parts, with two having opposite signs and therefore canceling each other out, and the third being positive. So the total must also be positive. If I had invested a little more time, however, and started my account by utilizing the Yang–Mills calculation rules (which I knew in theory but had never actually used), I would immediately have realized that it was necessary to add a fourth, negative part, which would dominate the result and render the sum negative as well. But I was pleased with the positive result, did not check it, and stuck with the mistaken belief. This is not the episode that I wanted to relate. It is a typical and not particularly significant error caused by haste. It is useful, however, for providing the background.

Soon after, the situation began to change rapidly. At a conference in Marseille in the summer of 1972, the Utrecht physicist Gerard 't Hooft, who was twenty-six at the time, announced that he had calculated the sign of the beta function in Yang–Mills and that the result . . . had a negative value! This grand announcement was met with indifference. There were few people present, and even they did not pay much attention. A friend of mine, an expert in the field, when questioned about it a year later, remembered that 't Hooft had said something, but he couldn't remember what exactly.

The only person really capable of understanding 't Hooft's result was Kurt Symanzik, a brilliant German physicist in his fifties, who urged him to write an article on the subject. Together with his thesis supervisor Tini Veltman, 't Hooft had just solved a fundamental problem with the theory of weak interaction (for which they shared the Nobel Prize in 1999) and had begun work on some extremely difficult calculations on quantum gravity. For him, the beta function calculation was hardly more than an exercise, and he did not have time to write it up for publication.

I happened to be good friends with Symanzik. In November that year I went to visit him in Hamburg for two weeks. He took me to a restaurant at the top of the television tower, where you could eat as much cake as you desired (they had six kinds, and I had a slice of each); we went to see an outstanding production of *The Magic Flute*; and he invited me to supper at his house to eat mackerel in oil with crackers, accompanied by ultra-pasteurized milk reinforced with condensed milk. We discussed physics for dozens of hours and dissected all the topics that we had a shared interest in. And yet, surprisingly, he did not talk to me about 't Hooft's result. Veltman told me a year later that Symanzik had said to him that I was "so wild"—so impetuous—that it was better not to mention anything to me, in case I might write an article using 't Hooft's result, though obviously not without acknowledging his contribution. It would have been perfectly legitimate for me to do so, but Symanzik wanted the result to be communicated to the world directly by 't Hooft, not by a third person acting as an intermediary.

It was only in February 1973 that I learned from Symanzik about 't Hooft's result. At the time I had just made important progress with phase transitions and did not give the result my full attention. But I had just moved to CERN in Geneva for two months, and because 't Hooft was working in the same research center, we decided to meet one morning to figure out how to use his result to build an asymptotic theory of the proton and other particles.

In fact we needed to identify the possible constituents that formed the basis of the theory and to verify that in that specific case 't Hooft's calculation gave a negative beta function. It sounded easy; in 1964 quarks had been hypothesized, and in 1971, Gell-

Mann, Bardeen, and Fritzsch had proposed the theory that each quark existed in three different "colors" that interacted by exchanging colored gluons—essentially the Yang–Mills theory studied by 't Hooft concerning gluons, with the addition of quarks. I knew Gell-Mann's theory perfectly: he had come to Rome and shown in a public lecture that the theory explained the data obtained by the ADONE accelerator at the Frascati laboratories, where I worked. Gell-Mann's argument was based on the hypothesis that quarks do not interact at high energies; hence the theory was asymptotically free. I had put my money on the opposite hypothesis—that quarks continue to interact even at high energies—and very presumptuously I had put down Gell-Mann's results as naive, inasmuch as they did not take into account all the complications of a theory in which quarks interact. Then I shelved it.

Looking back on it now, the conversation with 't Hooft was surreal.

"Ciao, Gerard, what a great result you've had. Let's see if we can use it to build a theory that describes the proton and other particles."

"It's a great idea! But how can we go about it? The Yang–Mills fields must have a charge of some kind. Which charge should we choose?"

"We could take the electric charge and other charges of that type."

"I don't think so, Giorgio. That would involve insurmountable difficulties with the experimental data."

"So let's try and see if we can find a loophole that will allow my proposal to work."

"No, actually it's impossible," he concluded, and gave a detailed explanation of the topic, which I could not fault.

"You're completely right! It's a real pity, but your theory can't be adapted to describe the proton and other particles. I'll see you over the next few days."

We did not give a moment's thought to the color charge proposed by Gell-Mann. It would have been enough at that moment to have seen his name written somewhere (on a blackboard, for instance), or for someone to have casually mentioned Gell-Mann's model at lunch or supper, for me to have been able to run to 't Hooft with a cry of "Eureka!" In a couple of days we would have done the checks, written it up, and sent it for publication. We were incredibly remiss, for which I take full responsibility. Gerard was a really profound theoretical physicist, capable of analyzing the most refined aspects of the theory. I, on the other hand, knew by heart the experimental work and the various models proposed in the literature. In other words, I was the one who should have identified the right model. On that morning in 1973 we let slip the chance to win a Nobel Prize. Fortunately, for both of us, it would not be our only chance.

A few months later Hugh David Politzer on the one hand and David Gross and Frank Wilczek on the other used 't Hooft's results at the same time and correctly identified the Yang–Mills fields. This marked the birth of quantum chromodynamics, and the article coauthored by the three scientists brought them the

Nobel Prize in 2004. Meanwhile I was left with a good anecdote to tell.

At a conference many years later, I met a friend of mine who had followed the story closely. While we were in the corridor we started talking about Kenneth Wilson, who had won the Nobel in 1982 for his theory of phase transitions. In particular we recalled Wilson's argument that a non–asymptotically free theory would be more elegant, but given that the universe was not designed by a tailor, the elegance of the theory was not a decisive factor. I added that at the time I agreed completely with Wilson, and therefore I too had not put much effort into finding a satisfactory theory that was asymptotically free. And I thought it was appropriate to tell my friend about that conversation I'd had with 't Hooft. He got to the point immediately, asking:

"But Giorgio, did it not occur to you to use color as proposed by Gell-Mann?"

"No."

"But how is that possible?"

"It just didn't occur to me."

"In that case, maybe you would have done better to have given it more than half an hour's thought . . ."

Another friend of mine told me the story of his grandfather who was a doctor and had noticed that certain molds secreted a substance that killed bacteria (what was called penicillin). This was a few years before Fleming, but unfortunately my friend's grandfather never wrote anything about it. He was convinced that it was a completely useless discovery: if the substance killed bacteria, which are extremely resistant, it must have been tremendously toxic to humans, so unusable as medicine. Antibiotics could have been discovered a few years earlier and who knows how many lives would have been saved.

Researchers often pass by great discoveries without being able to grasp them. You need great intuition to navigate infinite possibilities, and sometimes intuition fails.

Four

PHASE TRANSITIONS, OR COLLECTIVE PHENOMENA

Water boils, water freezes, and this is very strange indeed. We see a substance suddenly transform before our eyes, just because of a change in temperature. What we are witnessing is a collective mutation. It is not the single atom, it is not the single molecule of water that freezes or boils.

Phase transitions—abrupt changes in state due to changes in temperature—are phenomena of the "everyday physics" that we are so used to living with that we hardly notice them. But for physicists these are essential phenomena to try to understand. This is why, at the beginning of the 1970s, I was busy studying some types of phase transitions that at the time still represented an open question and a problem to be solved.

My superficial acquaintance with the theory of colored quarks was due to the fact that so much of my energy was devoted to studying phase transitions. In the spring of 1973 I made an important breakthrough in the field and was working hard to deduce all its implications. I was young and wanted to understand everything immediately. Over lunch at CERN, Tini Veltman advised me: "Don't do too many things—concentrate on a few important ones." On the one hand this was obviously good advice, but on the other it was precisely by studying many things at once that I was able to make connections between different fields—the basis of many of my later discoveries.

Phase transitions are phenomena that we are all familiar with, even if we have never heard this technical phrase for them. Everyone knows that when you heat water enough, it starts to boil. In other words, it begins to pass from a liquid to a gas, just as it passes from its liquid form to its solid one, ice, when made cold enough.

For physicists, the observation of these "normal" everyday phenomena generates countless questions. Why do such transformations occur? Why do they occur at these *particular* temperatures and not others? Do they happen in a similar way in all materials? And so on, in a series of puzzles for which it is very difficult to find answers.

In the first decade of the twentieth century, physicists began to see experimental evidence that allowed them to think of atoms and molecules as the "building blocks" of matter, and they tried to interpret macroscopic phenomena—such as the freezing of water—as emerging from the collective behavior of these extremely small units.

From a macroscopic point of view, phase transitions become

more difficult to describe, and represent a problem that returns in different forms. We began by resolving the simplest cases, and we have gradually refined the tools and increased the number and range of cases solved.

In order to study phase transitions at the microscopic level, we need to understand the behavior of many "objects," that is to say atoms or molecules or tiny magnets: those elementary things that—using a more general context than that of traditional physics—we can call "agents." These agents interact among themselves, exchanging information and modifying their behavior according to the information they receive.

In the context of physics, "exchanging information" is equivalent to "being subject to forces." But generally speaking—given that the model can be applied to many fields of study, from physics and biology to economics and so on—there are many objects whose behavior depends on the behavior of other objects that are more or less in proximity to them, given that objects that are too far apart from each other cannot exchange information.

The physical quantities that we can measure at a macroscopic level, such as the temperature of water, depend on the behavior of microscopic agents, for example the velocity of the molecules, which we cannot observe.

Let's imagine that we are looking at water through an extremely powerful microscope. We would see molecules in the shape of slightly bent dumbbells that move, are attracted by each other, rotate, are repelled, and vibrate rapidly. This is a description of water at the molecular level. If we look at it instead at the level of the human eye, we see a liquid that at a certain temperature freezes and solidifies and at another temperature evaporates, becoming a gas. How we pass from the behavior of single atoms to the global behavior of the system is a problem that has taken some time to explain.

THOSE WHO STUDY phase transitions are not so much interested in understanding at what temperature a certain change of state happens as in the mechanism that causes it to do so. Why is it, for example, that this phenomenon happens all at once, at a point that is so specific? What changes within the system at one hundred degrees Celsius? Why is it that if we observe the system at just one degree below the critical temperature, we notice nothing? And why is just one degree extra sufficient to bring about such a sudden macroscopic change?

Solving this problem conceptually is by no means an easy or trivial matter—indeed, in the 1930s, many physicists were asking whether the normal rules of physics, and in particular those of quantum mechanics, would be sufficient to explain phase transitions.

The solution was found in the 1940s and '50s, starting from an idea that was already well known in physics: energy minimization. In nature, an object that is free to move will seek its position of lowest energy, until a point of equilibrium is reached. A ball set rolling downhill, for example, will come to rest at the lowest point of the incline. The position it has at the bottom represents a stable equilibrium position, and the ball will stay there unless something else intervenes to cause it to move again.

Something similar happens with ice. At a temperature of zero degrees Celsius, it finds itself in a stable equilibrium state (solid) that corresponds to a minimum of its free energy. With an increase in temperature, the molecules—which in the solid phase occupy precise positions in the crystal lattice—begin to agitate until they lose their fixed positions and move freely. This is the liquid phase, and it too represents a stable equilibrium corresponding to another minimum point of free energy.

Providing heat to the water is like kicking the ball: however lightly it is touched, the ball will start to roll. But it may not have enough energy to leave the bottom of the incline, and it will move only until it finds another equilibrium position.

Just so, the molecules in the crystal lattice that defines the solid phase increase their agitation as the temperature increases. When the system reaches zero degrees, the links holding the molecules together begin to break. In this phase, continuing to supply heat does not increase the temperature but gives the system the energy needed to break the links between molecules, until all the ice has melted into water, a new stable equilibrium position in the liquid phase.

This type of phase transition, called "first order," is characterized by two important phenomena.

The first is that the system, when close to the critical point, does not show any microscopic characteristics that would lead us to think a transformation was imminent. Water at 0.5 degrees Celsius does not show any indication that would allow us to understand that it would begin to freeze if we lowered the temperature by half a degree. No islands of ice form in the water, or islands of water in the ice, when the system nears the critical temperature.

The second significant phenomenon is the existence of *latent heat*, that quantity of heat needed to break the molecular bonds rather than increase the temperature of the system. The heat we provide when the ice is at zero degrees will serve to break the bonds until all the ice is melted. This quantity of heat that the system needs in order to change state is called latent heat.

Phase changes in this direction can sometimes be described as transitions from an ordered to a disordered system state. In the solid phase, molecules occupy precise points in the crystal lattice and are therefore in an ordered phase. In the liquid phase, the molecules of water are able to move freely and the microscopic situation appears much more disordered than in the preceding phase.

NOT ALL MATERIALS BEHAVE like water does. There are other phase transitions that happen without latent heat—that is to say, without needing a certain quantity of heat once the critical temperature has been reached in order to pass from one state to another. In such cases the transition occurs continuously, we might say smoothly, as the critical temperature is approached. These are called second-order phase transitions.

Let's take an example: a magnet, which at room temperature is a magnetic system, loses its magnetization as the temperature increases. In technical terms we say that it passes from a ferromagnetic (magnetic) to a paramagnetic (nonmagnetic) phase. It is possible to visualize the magnetic field of the magnet as an arrow oriented in space, exactly like the needle of a compass, with the tip of the arrow pointing north.

This macroscopic magnetic field is generated by the sum of the great many elementary magnetic fields of every particle in the system, which we call *spin*. Inside the magnet, the interactions that exist between the spins cause them to align: a very large number of small arrows all pointing in the same direction.

In the case of magnetization, the phase transition also happens with an increase in temperature. The heat supplied to the magnet causes an increase in spin movement that can change the orientations of the spins. They tend to become disordered and lose their alignment. And it is precisely this alignment of spin that generates the macroscopic field, which, as temperature increases, diminishes until it is completely nullified.

In this case too, then, the phase transition can be described as a transition between a more ordered and a less ordered phase of the system.

To illustrate this, we can use the model proposed by Ernst Ising in his 1924 doctoral thesis—perhaps the first model invented by physicists to aid an understanding of reality by simplifying its description as much as possible. This model allows spins to orient themselves in just two directions: up or down (as shown in figure 1), with all other orientations forbidden.

The force that exists between the spins is such that they will tend to line up in the same direction (all pointing up or all pointing down), though thermal agitation will cause some to be reversed and go in the opposite direction.

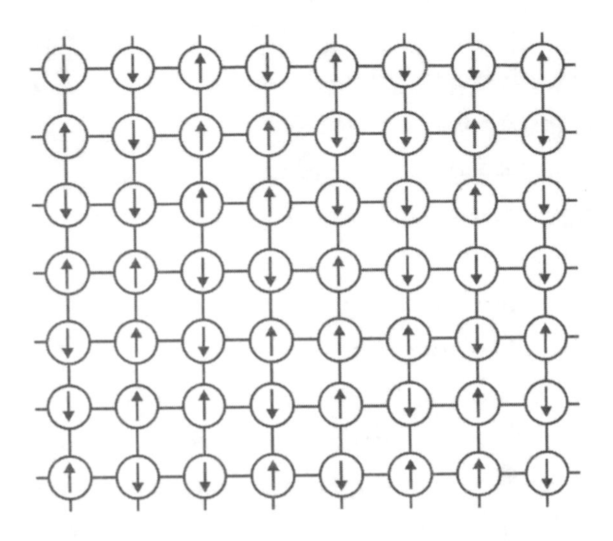

Figure 1. An Ising lattice, with spin represented by arrows that may point in different directions.

The ferromagnetic phase will correspond to the majority of the spins being oriented in the same direction (ordered phase), whereas the paramagnetic phase will be described as 50 percent of the spins pointing up and 50 percent down, randomly distributed throughout the system (disordered phase).

The system can also be described in terms of symmetry. If a transformation does not change the system's characteristics, we can say that it is a symmetry of the system.

Let's take as an example the transformation "inversion of all spins." Applying it to the disordered (or paramagnetic) phase changes nothing, because we still have 50 percent of the spins pointing up and 50 percent down, always randomly distributed. This is a symmetry of the system.

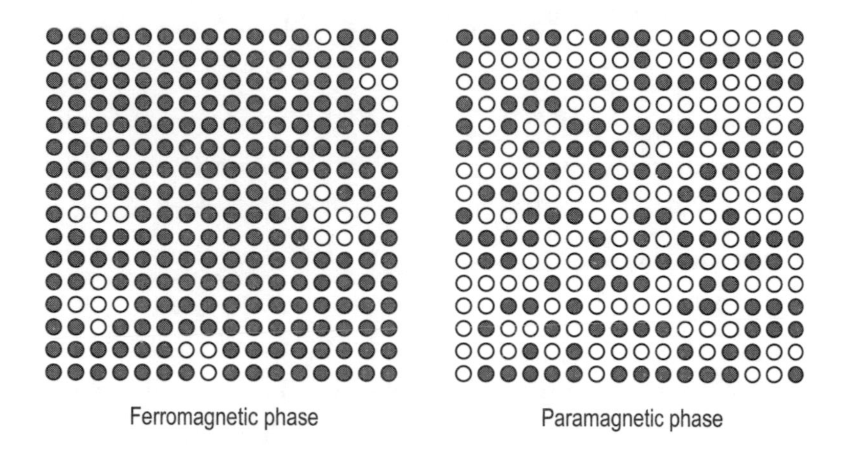

Figures 2a and 2b. Phases of the Ising model. The color gray indicates spins that are oriented down; white shows the spins pointing up. In the ferromagnetic phase, we see small islands of spins that point up (white), whereas the others, representing the majority (gray), point down. In the paramagnetic phase, the spins are distributed randomly, half up and half down.

Below the critical temperature, on the other hand, when most spins point in the same direction (as in figure 2a, in which the majority of the dots are gray), their inversion results in the inversion of the macroscopic magnetic field, which will change its sign (that is, the majority of the dots will become white). For the ordered (or ferromagnetic) phase, spin reversal is not invariant, because it actually reverses the magnetic field.

In this case we say that a "spontaneous breaking of symmetry" between the two phases has occurred: a symmetry (inversion of spin) in the paramagnetic phase no longer exists after the phase transition, when the system finds itself in the ferromagnetic phase.

This symmetry is spontaneously broken, without the intervention of external phenomena.

Magnetic transitions form part of the class of second-order phase transitions, the ones characterized by a parameter, in this case magnetization, called the "order parameter," which describes the passage between an ordered phase of the system and a disordered one.

At first glance, the magnetic system seems similar to that of water, because there are no discontinuities. But the devil is in the details, and the details of second-order transitions are very complicated.

Let's take a magnet at a temperature high enough for it to exhibit no magnetization whatsoever and place it in a magnetic field before gradually lowering its temperature. We will see that the system becomes magnetized more and more the closer it gets to the critical temperature. Once reached, the transition occurs and the magnet acquires its own magnetization, without any need for an external field.

Exactly at the phase transition point, we have large ferromagnetic regions: in some of the regions most of the spins are up, and in others the spins are down. This situation is very complicated to study, and is illustrated in figure 3.

HERE IS AN INTERESTING FACT that has been measured by experimental physicists: the behavior of a magnetic system does not depend much on the behavior of the individual elementary objects of which it is composed.

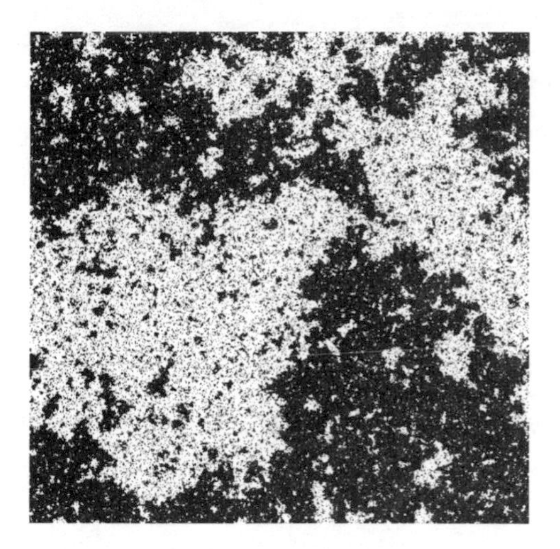

Figure 3. A one-megapixel picture of the two-dimensional Ising model at the critical temperature: each pixel represents a spin that can be black or white (up or down). One can see the large regions where the spins are predominantly black or white.

By comparing extremely different magnetic substances in which the interactions between the microscopic components and their description of quantum details are different, it has been observed that the magnetization diminishing to zero in proximity to the critical temperature always follows the same trend. This trend is mathematically described by a power-law function that presents similar characteristics for a whole class of magnetic substances, including ones that are very different from each other.

It is as if the cars in a Formula 1 Grand Prix did as they liked for the entire race, but then all slowed down in the same way on the last lap in order to stop on the finish line.

This is a surprising, unexpected discovery: despite the microscopic details being completely different, the collective behavior was the same. This result was formalized by American physicist Leo Kadanoff, who coined the idea of *universality classes* into which phase transition phenomena could be divided. Phenomena with the same beta exponent value belong in the same class.

It's a fact that recalls the Platonic view of nature: it could be said that there are relatively few universality classes of critical behaviors, and each actual system leads back to one of these universality classes—in other words, in Plato's terms, to an Idea.

The classes are divided based on the degrees of freedom of the systems' elementary components. For example, particles that can move in all three dimensions of space, versus ones that are confined to a single plane, have different degrees of freedom. In short, a material's universality class depends on how, and how much, the elementary constituents can move, and the value of the beta number depends solely on these degrees of freedom.

In the early 1970s, this problem—a concrete example of which we will look at—was rightly considered to be a very interesting one, and the feeling was that all the tools existed to solve it if we could find the appropriate formalism to calculate the critical exponents. And so I started working on phase transitions, thinking that in a short time I would arrive at a solution and then go back to more difficult-seeming open problems concerning elementary particles.

WE WERE ESSENTIALLY STUDYING systems with strong magnetic interactions between spins. We understood these interactions at a microscopic level. We wanted to find a formalism that, starting from this known microscopic description, would be able to describe the system at an intermediate level without referring to microscopic details since the behavior of magnetization does not depend on them. At this intermediate, or *mesoscopic*, level, we study the fluctuations of the system: groups of more or less numerous atoms passing from one phase to another. The evolution of the system can be analyzed by studying these fluctuations and their reciprocal interactions. As we will shortly see, the fluctuations are independent of the scale used to analyze the system.

There had been a great deal of work done, such as that by Giovanni Jona-Lasinio and Carlo Di Castro, two well-known theoretical physicists from Rome, to try to understand in detail the origin of mesoscopic behavior. Two fundamental advances were made by Kenneth Wilson, in articles published in 1971 and 1972, on how to build a formalism that would allow the calculation of critical exponents. This formalism, dubbed renormalization group (RG), earned him the Nobel Prize in 1982.

To understand why the technique proposed by Wilson for dealing with second-order phase transitions was given the name "renormalization group," we need to have a general idea of the procedure he used. The description of the system at an intermediate level is an invariant description for a transformation of scale: in other words, the result of our observation does not depend on how much we use the zoom.

Take a look at figure 4.

The image on the right is an enlargement of that part of the image on the left contained within the box. As the figure shows,

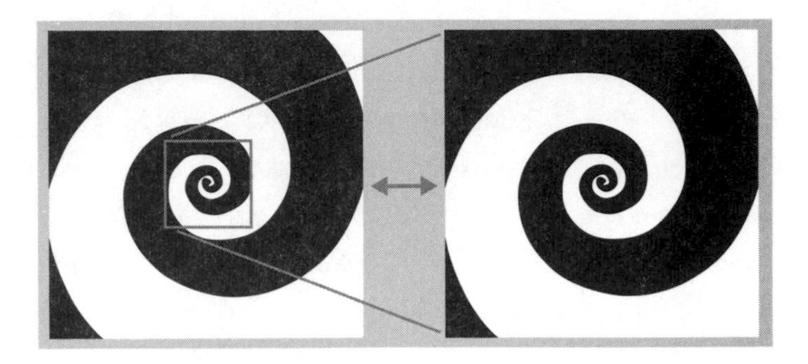

Figure 4. Invariance of scale for a special kind of fractal figure. By zooming in on the region inside the square on the left, we find the original figure again.

there is no way of distinguishing between them when the observation scale varies—or, if you prefer, according to the zoom that we are using to look at them.

Let us now turn to our system as illustrated in figure 3. Its fluctuations behave in essentially the same way, apart from a scale factor: the farther away we are from the system (we can think of using a wide-angle lens), the smaller the fluctuations will appear. And the closer we get (with a zoom), the bigger they will appear.

The idea, already introduced by Kadanoff, is to divide the system into squares that each contain a certain number of spins. Look at figure 5a: each 3×3 square groups together nine spins. The next step is to count how many of these nine spins point up (black) and how many point down (white). If we take the 3×3 square in the top left-hand corner, we can see that it contains six small black and three small white squares. Then we calculate the top left-hand corner of figure 5b as if it were a single entity, a

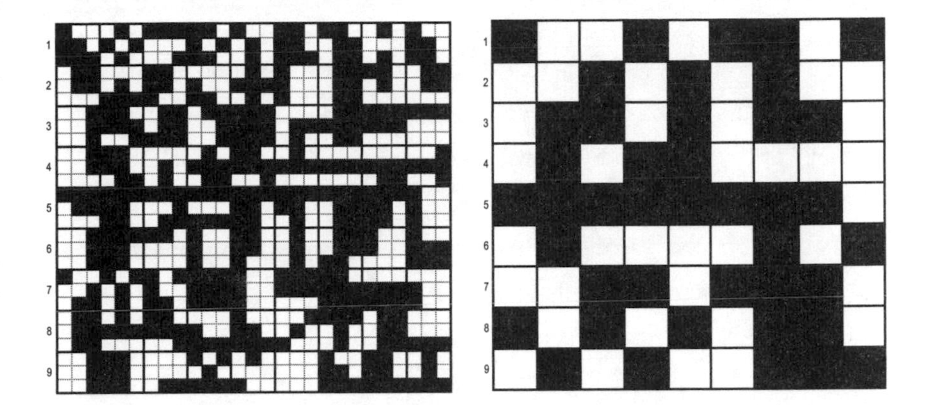

Figures 5a and 5b. Figure 5b was constructed by taking groups of 3×3 squares from figure 5a and coloring the corresponding square in 5b black, if the majority of the nine squares was black, or white, if the majority was white.

single spin. Each square in figure 5b is formed by determining the majority, black or white, of the nine spins in the corresponding area of figure 5a. So, for example, the square in the top left-hand corner of figure 5b is effectively black. In other words, we use a system analogous to the one used in American presidential elections: the candidate who achieves a majority in a state gets all the delegates in that state.

Every time we do this, we are actually changing the scale and significantly reducing the number of variables to be taken into consideration: instead of the nine spins in the top left of figure 5a, we have a single spin in 5b. This new representation of our system at the larger scale is still a good one: we are just looking at a "grainier" image. Wilson's technique allows us to pass from one scale to the next, and from this we get the term "renormalization."

I too, influenced by the Roman school of physics, had begun trying to figure out how to calculate the critical exponents, just as Wilson's works were about to arrive. I remember meeting him at a conference in Rome toward the end of 1971, before his decisive work in '72. After the conference, I said to him: "The information on the value of the critical exponents is in the first Feynman diagrams," and he replied cryptically: "I know that; the problem is how to extract it." The crucial work showing how it should be done came out a few months later. For less than a year I saw him as a competitor. But there was really no competition; he was better and was completing the final part of a race after years of preparation. I consoled myself by introducing an important improvement to the Wilson calculation of the critical exponents, and that was what kept me busy during that fateful beginning of 1973.

By the end of that year, I thought I understood the basic principles of these calculations. The phase transitions of magnetic systems had found one of their deepest scientific explanations. Satisfied with my contribution, I turned my attention to the physics of elementary particles, where it seemed to me that there were still many important problems to be solved.

Five

SPIN GLASSES: THE INTRODUCTION OF DISORDER

Often the best work achieved in a life spent in research happens by accident. You start out on one road only to arrive at a quite unexpected destination.

This is what happened to me. The theory of spin glasses, considered my most significant contribution to physics, started while I was working on a problem to do with elementary particles.

It seemed to me that the tool best adapted to solving that problem was a certain mathematical technique I was not familiar with yet, called the replica method. I got hold of all the existing literature on the method and started to study it. With the replica method, you take a system and mathematically replicate it several times, then compare the behavior of the different replicas. It seemed very suitable for solving my problem, but in one of the cases described in the literature, it yielded results that were inexplicably incoherent.

Confronting a problem that is new and therefore by definition unclear, with a tool that might or might not work, is not such a great idea. It is like using a compass that occasionally points south instead of north, without anyone knowing when this might happen or why.

I decided to find out for myself how reliable this instrument was. It was just before Christmas 1978. I photocopied the article detailing the case in which the replica method yielded unreliable results and took it with me on holiday.

The article focuses on problems related to disordered systems and spin glasses, topics I had never dealt with and that were remote from my own field of study at the time. I had to understand why the method had not worked in this case. I studied the model and redid all the math—the math was right, but the result was incongruent. Clearly I had to look into it more deeply.

When I came back from holiday, I found certain works that were showing progress, and the solution seemed close at hand. I tried to solve the problem starting from those more advanced studies, thinking that it would be relatively easy to achieve—but the more I worked on it, the more difficult it seemed to become.

If some results became consistent, others deviated from the values of the numerical simulations—an indication that the solution was far from close. It was likely that a radical change of perspective was needed.

Without even noticing, I was exploring a new field of research. I was no longer thinking about the elementary particle problem with which I had begun. My interest had been drawn to something else entirely.

SPIN GLASSES ARE METAL ALLOYS, so called because their magnetic phase transitions, due to the spin behavior of the particles that form the alloy, behave like the phase transitions of glass.

These alloys are composed of noble metals, such as gold and silver, within which a small amount of iron has been diluted. At high temperatures they behave like normal magnetic systems, but when the temperature falls below a certain value, they appear to behave like glass, wax, or bitumen: the changes get slower and slower, and it seems as if the system never reaches equilibrium.

At school we studied how a liquid is a material that takes the form of the solid container into which it is poured. At a high enough temperature, glass is a liquid, but it behaves in an unusual way. If we take a container full of molten glass (or honey, or wax) and turn it upside down, the liquid does not immediately fall to the ground, but instead begins to slowly ooze from its container. The more the glass cools, the slower it pours: for whatever reason, the behavior of the system slows significantly.

The very strong slowdown in the dynamics of the system as the temperature lowers has something in common with the magnetization behavior of metal alloys. It is as if by lowering the temperature, we diminish the ability of the spins to move, making it impossible for them to reach an equilibrium position.

Let's go back to our example of the bus full of people from chapter 1: as long as the density is relatively low, a person can get past the others and go from one place to another inside the bus. Anyone who moves will make others move, in a causal chain. Everything works if there is enough space, but the higher the density, the more immediate the contact between people becomes: space between passengers decreases; it gets increasingly difficult to maneuver, and people become stuck, like in a traffic jam.

The phenomenon is general enough—it applies to glass, wax, honey, pitch, metal alloys—to have attracted research and investigations into how it works. The best way of approaching it is to build a model, a simple one at first, that reproduces the phenomenon. This enables us to find the essential characteristics or interactions that lead to a slowing down of the dynamics with a variation in temperature. Characteristics and interactions that, though present in glass, honey, wax, bitumen, and some metal alloys, are absent from water and almost all other liquids, which do not exhibit this behavior.

STUDY OF THE PHASE TRANSITIONS of these materials is difficult from an experimental point of view. In Australia, researchers have been conducting a unique experiment for decades. A funnel of pitch is kept at a controlled temperature where it retains some viscosity—it can move and drip from the funnel—and the frequency with which the drops fall is measured. The experiment was begun in 1927. As of 2022, a grand total of just nine drops had fallen.

These systems are complicated to study, and the best way of doing so is to build a synthetic model that is simpler than reality, which can help with finding solutions and understanding the systems' characteristics. To understand what a model is and why it's useful for a theoretical physicist, we can think about the game of Monopoly. It is a model of society with only a few simple rules: the location and cost of land, the cost of building on it, and the value of real estate rents. Random elements of the kind that are always present in our lives are then added: a roll of the dice to move, the "unexpected" and the "probability" of escaping from or entering into difficult situations.

With these simple rules, the game reflects a fundamental characteristic of capitalist systems: the more money you have, the more you make and the richer you become.

Just as Monopoly does not contain all the complexity of a real society but nevertheless manages to reflect some of its characteristics, so the models built by physicists do not contain all the complexity of real systems—and yet, if we introduce significant rules into the model, we can hope that they will reproduce some of the fundamental characteristics of the phenomenon that we want to study.

Once the model has been built and we have inserted the rules that describe its operation, we can begin to evolve the system. We can start our game of Monopoly, or simulate on the computer the phase transition of our system, raising or lowering the temperature of our synthetic model.

In evolving, the model will generate certain results: in the case of Monopoly, "whoever has the most money will get richer and richer"; in the case of Ising's model, the ferromagnetic phase that emerges when the temperature drops.

Then we begin the work of developing the theory, or the mathematical structure that reproduces the results of the simulations, starting from the rules and initial data of our synthetic model. An experimental laboratory no longer consists of magnets, circuits, ovens, and the like: it is now our computer.

If we can succeed in doing this, we will worry later about how the theory can be used in actual cases—for metal alloys, glass, wax, and numerous other systems.

IN THE ISING MODEL of ferromagnetism that we looked at in the previous chapter, the forces between spins are such that at low temperatures they tend to align in the same direction, all pointing up or all pointing down. In the spin-glass model, on the other hand, the force acting between some spin couples tends to orient them in opposing directions, and this complicates the situation.

Let's consider a practical example of this. In our lives our own objectives are frequently at odds with the objectives of others, forcing us to not pursue them. I want to be friends with Mr. White and Mr. Green, for example, but unfortunately they happen to detest each other, making it difficult for me to be very close to them both at once. This situation, frustrating enough in itself, becomes even more complicated when many people are involved.

Let's imagine a tragedy of this type: there is a conflict between two groups, and every individual must decide which side to be on. Each person, moreover, has strong feelings of either sympathy or antipathy toward each of the others (It really is a tragedy . . .).

For the sake of simplicity, we can assume that all feelings of sympathy and antipathy are mutual.

Now let's take three participants in the drama: Anna, Beatrice, and Carlo. If all three are friendly with each other, there is no problem: they will simply choose to be in the same group. The outcome will be equally simple if two of them are friends and both feel dislike, which is reciprocated, for the third. In this case the two friends will choose one group and the third person the other one. But what happens if all three individuals dislike each other? The outcome is an inevitable degree of frustration, because two of these individuals who mutually dislike each other will find themselves having to share the same group.

When numerous groups of three are similarly frustrated, the situation begins to become unstable, with some individuals perhaps changing groups in search of a situation in which the total amount of frustration is lower. It is possible to define the degree of "dramatic tension" by dividing the number of frustrated groups of three by the total number of trios.

Detailed studies have shown how in a Shakespeare tragedy the dramatic tension defined in this way is quite low at the beginning of the play, reaches a maximum at around its midpoint, and then decreases toward the end.

In the spin-glass diagram in figure 6, in which we are no longer dealing with trios but with spins positioned on a square grid, each spin can point only upward or downward (any other orientation is prohibited). What we could previously define as "sympathetic bonding" we now call "ferromagnetic bonding"—a force that tends to align the spins in the same direction, which we have represented

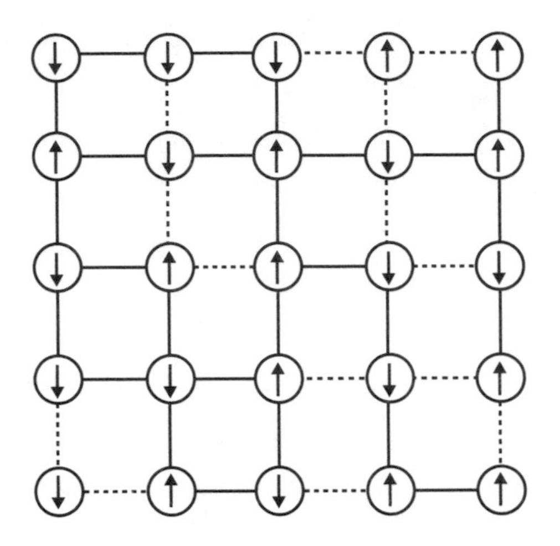

Figure 6. A representation of spin glasses on a plane. At low temperatures, the spins connected by the dotted lines seek to settle in opposite directions, while the spins connected by the solid lines orient in the same direction.

in figure 6 with solid lines. Bonding through antipathy becomes "antiferromagnetic bonding"—indicated by the dotted lines and referring to a force that tends to align the spins in opposite directions. In this case too it is possible to easily identify situations in which there is frustration. Look at figure 7, for example.

In this case, the spin in the top left-hand corner has an antiferromagnetic coupling with the spin below it and a ferromagnetic coupling with the spin to its right; therefore it can satisfy only one of the two and does not know whether to align itself upward or downward.

The first spin-glass models were used by Sam Edwards and

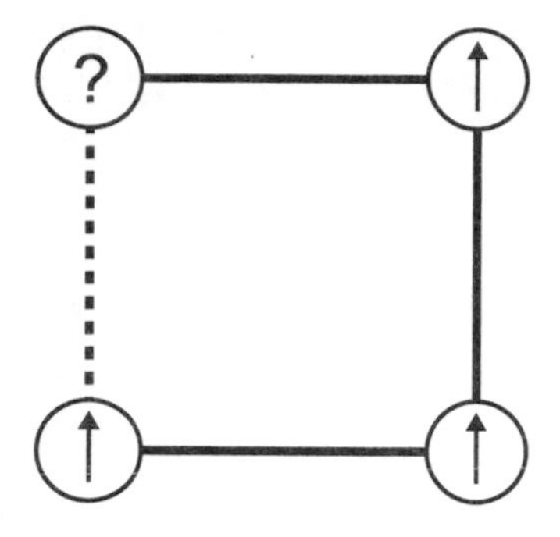

Figure 7. The three links represented by the solid lines are ferromagnetic, whereas the one represented by the dotted line is antiferromagnetic. The spin depicted with *?* has no reason to point up or down.

Philip Warren Anderson, but a simpler model was built by David Sherrington and Scott Kirkpatrick in 1975.

Returning to our problem, if we used the replica technique to calculate the physical quantities of the system of spin glasses described by the Sherrington-Kirkpatrick model, we arrived at a series of inconsistencies. The computation of entropy, for example, led to negative values—which is impossible, given that in any physical system, entropy is by definition positive. If the calculation of the entropy of a system leads to a negative result, either the calculations are wrong (which can happen, though our checks showed that this was not the case) or somewhere along the line there has been a serious conceptual error.

INDEED THERE WAS A conceptual error. It was technical, and as such difficult to explain in lay terms. Suffice to say that it was linked to erroneous mathematical assumptions that greatly simplified the problem. Someone had already discovered the error, but no one knew how to solve the problem without these simplifying assumptions, which were wrong.

In the first work I wrote on the subject, in 1979, I showed that it was possible to use a given construction to partially solve the problem. At the end of this paper, I added sanguinely that "this construction can be generalized in order to arrive at a complete solution."

As is the case with all scientific articles, before publication it was sent to be peer-reviewed, meaning read by a colleague in a position to judge whether it merited publication or not. His verdict was more or less that what I was doing was "incomprehensible." However, "given that the equations yield results in accordance with the numerical simulations, the work can be published. As for the part relating to the generalization of the approach to more complicated cases, it is not worth the paper it is written on." So the article was published after all, but with the final part of it cut.

Joking aside, the truth is that I did not myself really know what I was doing. I had found some rules with which to tackle the problem, I had applied them, and in the end, after a series of steps, I had come up with equations that were meaningful and fundamentally reproduced the data of numerical simulations, giving a value for entropy that was now positive.

But I had no idea what had happened in the course of the calculation. It was as if I had gone into a tunnel and then found myself on the other side.

In the article that followed, the agreement between the results provided by the theory and by simulations suggested that the theory might make sense, though what that sense was still remained obscure.

The physical fact that I did not understand was related to what physicists call an order parameter. The changes of state in a system, as we have seen, are generally characterized by the variation of a parameter. When studying the transition between liquid and gas, for example, the order parameter is density. In the case of magnetic transitions, the order parameter is magnetization. These order parameters vary during phase transitions, assuming different numerical values, the physical significance of which—as with density and magnetization—is very easy to understand.

Surprisingly, however, in the case of the results of my calculations for spin glasses, the order parameter was not a simple number that changed during the transition. What changed during the transition was a function. A point was not enough to characterize the transition; instead I had to use a function consisting not of a single number but of infinite numbers.

What did this function represent in physical terms? To have a function instead of a number as an order parameter for a transition also indicated a watershed for employing the replica method. In cases in which the parameter is only a number, the replica method gives absurd results, whereas when the order parameter is a function—that is, an infinite set of numbers (in the same way that a line can be seen as an infinite set of points)—then the replica method produces results that are coherent.

Clearly there had to be some deep physical significance linked to the need for an infinite number of parameters (i.e., for a function) to describe the transition of a system, but at this time that significance was a complete mystery to me.

BEFORE GETTING TO THE PHYSICS, let's try to understand the modification that was needed from a mathematical point of view.

To make the replica method work, I had to "extend" it. The possibility of extending a mathematical method is based on a very old idea. The French bishop, mathematician, physicist, and economist Nicola d'Oresme was probably the first to use it, in the mid-fourteenth century. D'Oresme was an incredible figure, and provides conclusive proof, if proof were needed, that the medieval period was not the dark age for science that our school textbooks might have us believe. To choose just one of many examples, he was the author of a book on how the positions of stars are distorted by atmospheric refraction, written around 1360. Since it is in Latin, I cannot claim to have read it through. It is nevertheless obvious that his reasoning was remarkably sound from a conceptual point of view. He probably got the idea while observing the sun "crushing" the horizon at sunset, suggesting that there must be a distortion. Calculating the degree of distortion is vital if we are to make precise astronomical observations, as the apparent measurement of the stars needs to be corrected by as much as two or three degrees.

Returning to our problem, d'Oresme was the first to realize

that raising a number by ½ was equivalent to finding its square root. This seems now like a banal enough truth, of the kind that we all learn in high school—but this fails to recognize the logical leap that d'Oresme made by extending the properties of powers to fractional numbers. These were properties that had been reserved exclusively for integers.

The idea of raising a number to a power is very simple. Squaring a number, or raising it to the power of two, means multiplying the number by itself. Cubing a number, or raising it to the power of three, means multiplying the number by itself, then by itself again. And so on. To raise to the power of ½ seems absurd: What does "multiplying a number half times" mean? D'Oresme's idea was to extend a property of exponentiation: that exponents must be multiplied when raising to a power a number that has been raised to a power. For example, 2^2 raised to the power of 3 is equivalent to 2^6 (i.e., 64, or 4^3).

If squaring a number raised to the power of ½ results in the number with which we began (since 2 times ½ is 1), then raising to the power of ½ is equivalent to extracting the square root: the square root of a number squared is that number itself.

These properties are formally derived, since multiplying a number half times makes no sense; the formal properties, however, guarantee a coherent result. Nicola d'Oresme went beyond the original point of view, beyond immediate understanding, but by maintaining the formal properties, he obtained a very simple method for solving even complex operations. After d'Oresme, mathematics often advanced by extending properties in a formally correct way in new conditions, broadening their scope.

To solve my problem, I used a similar method. I formally applied mathematical techniques developed and verified purely for integers in the hope that the properties of the formalism would remain valid for non-integers.

The idea I had was an extension of combinatorial calculus. Combinatorial calculus tells me, for instance, how many ways there are to arrange ten objects in pairs in five drawers. By extension it is possible to use the same equation to find how many ways five objects can be arranged in ten drawers, meaning each drawer contains an object "half the time." Obviously the result does not quite make sense, because it cannot be reproduced in practice: we cannot resort to cutting the objects in half, yet it amounts to saying that the number of objects in each drawer is ½. But in order to obtain a normal solution relating to actual things, it was necessary to go by way of these imaginary objects: drawers that contain an object half the time, a non-integer total number of objects, and a non-integer total number of ways in which we can put fractional things in drawers!

Starting from this procedure, my idea was to divide the objects in half and then in half again, and then again and again, reducing the objects in the drawers toward zero. This is an exclusively mathematical procedure with little physical sense—but it led to correct results that were compatible with data from simulations.

Two open questions remained: how to demonstrate that mathematically it made sense to perform such an operation, and how to understand the physical significance of an order parameter described by a function and not given by a single variable.

. . .

AFTER A FEW YEARS the mathematical language of the replica method was translated into the language of statistical physics, becoming much more understandable even if the formulation of it was much more verbose. I made the fundamental breakthrough together with my friends Marc Mézard, Nicolas Sourlas, Gérard Toulouse, and Miguel Virasoro, who at the time were all based in Paris.

Using a series of indications, or clues, we managed to understand the physical meaning of the result, a common characteristic of all disordered systems. Namely, that disordered systems are simultaneously in a very high number of different states of equilibrium. It was a totally unexpected discovery.

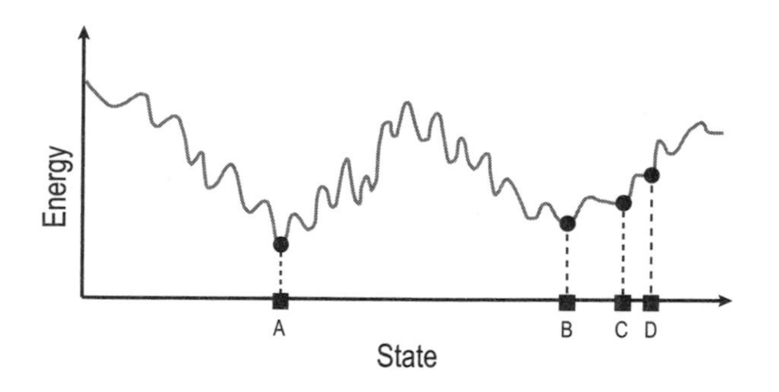

Figure 8. At low temperatures the system moves toward regions of low energy, and can be in any of several states represented by the line.

As we can see in figure 8, the system can be in any of the states found along the drawn line. (It can be, for example, at any of the points marked from A to D, which represent just four of the very many possibilities.) The states of the system have different energies, and there are many energy minima (holes) where the system reaches equilibrium. In the state marked A, the system is at the lowest point in that zone, as it also is at B—whereas in states C and D the system is in a shallow dip (that is, a situation of equilibrium from which it will not escape without the temperature of the system being raised), but a dip that does not represent a minimum for the zone.

In the diagram there are also two wide valleys (the region around A and the region around B) within each of which there are many small depressions, or lows. Let's call them Zone M and Zone N (figure 9). When the system, cooling down, ends up in a state in Zone N (at B, C, or D, for example), it will tend to remain in this zone even as the temperature rises, as long as the increase is not too high. The system evolves within a zone, that is, within a set of configurations that have been selected from the history of the system, or rather from the zone, from among the very many possibilities that the system has found itself in when the temperature drops.

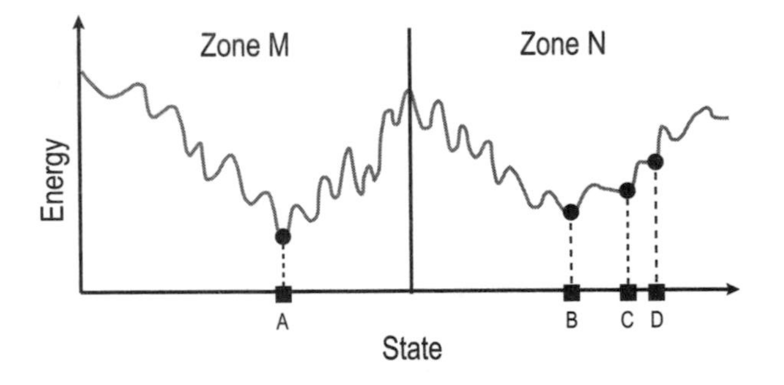

Figure 9. Two wide and deep zones in which the system might evolve.

Normally a physical system is found in a single state. Water, for example, at a certain temperature and a certain pressure, is either a solid, liquid, or gas. There are particular cases in which the system can be in two states, or phases. At the boiling point, water can be simultaneously in the liquid and gaseous phases. There is also a unique pressure and temperature value at which water can be in all three phases. This is the famous "triple point" of water, and it is famous for a reason. In general a system is in only one phase. A disordered system at low temperature, on the other hand, is in a very large number of phases at the same time. This is the meaning of the order parameter becoming a function, or an infinite set of values.

Understanding this represented a real step forward for physics. The construction of a synthetic model and its solution allowed the discovery of a new phenomenon and opened wide a door into the world of disordered systems.

Starting from the physical interpretation, we have succeeded in arriving at a mathematical one. It took more than twenty years to do so: it was obtained by Michel Talagrand, and the prior work of Francesco Guerra and his collaborators was fundamental in finding the key to the problem. The arguments used in the proof are really ingenious in their simplicity. But then everything seems simple with hindsight.

THE SOLUTION FOUND FOR spin glasses is a good starting point for studying actual glass of the kind found in windows, the behavior of which is not fully understood by physics. Attempting to come up with a description that allows us to understand all aspects of glass transitions is something I have been working on, on and off, since the middle of the 1990s.

Like spin glasses, real glass exists within a disordered system; this disorder is due to the fact that glass is made up not just of silicon but of many impurities, many different types of molecules of different sizes, all mixed up together. Therefore glass is not able to crystallize, because crystallization requires regularity of structure. The disorder of metal alloys called spin glasses, as we have seen, is due to the causality of the arrangement of iron atoms within gold: when the metal is liquid, the iron atoms can move randomly in the gold, but when the alloy cools, their movement is decreased until they are trapped in random fixed positions.

When trying to get a concrete understanding of real processes, everything seems incredibly complicated—but as soon as the work is completed, it seems beautifully simple. When we study a physics theory or a mathematical theorem in a textbook, everything seems perfectly clear. The amount of complicated work that was necessary to obtain the result has completely disappeared.

Another interesting problem to be tackled is transitioning from a schematic model, such as the spin-glass model we have just illustrated, to a more realistic one in which the forces between the spins are described in more detail—by taking into account, for example, the distances between spins. The phase transition occurs through interactions between objects that involve precise spatial dislocations, something that is not factored into the simplified model discussed above.

In addition, the simple model lacks consideration of evolution over time. The techniques of statistical mechanics can be "easily" used when a system is in equilibrium, that is, when it remains stable over time. For a disordered system such as glass or wax, the amount of time needed to reach a condition of equilibrium is generally very great. It can take years, even centuries to happen. This is also the case for the glass in our windows, which is treated with certain industrial techniques to make it more rigid.

Whenever a physical process is not in a state of equilibrium, time enters into the equation. It is always possible to distinguish a before and an after, which we can't do with systems in equilibrium. To simplify, if we find a ball in a state of stable equilibrium, that is, at rest at the bottom of a depression, and take photos of it, we will never be able to put those photos in the chronological order in which they were taken, because there will not be a single difference between them, no sign that anything has changed. The situation is transformed if we take photos of a ball rolling down an incline. In a state of nonequilibrium, the temporal sequence is not only evident but frequently very obvious.

We are therefore faced with the problem of extending the theory to incorporate time, given a situation of nonequilibrium, as well as a spatial dimension, given that the processes occur in space and the interactions are only between neighboring particles. In short, there is still a lot of work to do in order to fully understand transitions in glass.

I STARTED BY UTILIZING a mathematical technique in order to solve a particle problem (and for this problem, the replica technique in its original version worked superbly well) and ended up with a mathematical and conceptual tool that was very powerful and useful for solving a wide range of apparently disconnected problems. Namely, problems having to do with disordered systems.

The actual world is disordered, and as we said at the start, many situations in the real world can be described as a large number of elementary agents that interact with each other. These interactions can be schematized with simple rules, but the results of their collective action are sometimes really unpredictable. The elementary agents can be spins, atoms or molecules, neurons, cells in general—but also websites, financial traders, stocks and shares, people, animals, components of ecosystems, and so on.

Not all interactions between elementary agents generate disordered systems. Disorder is born from the fact that certain elementary entities behave differently from others: some spins try to go in opposite directions; certain atoms are different from most others; certain financial actors sell shares that others are buying; some dinner guests actively dislike others who have been invited and want to sit as far away from them as possible.

In all these disordered cases, the mathematical and conceptual tool I found is indispensable for tackling the problems associated with them. Recently, for example, we have achieved important results while trying to solve the problem of putting into a box as many different-sized solid spheres as possible. It's a very interesting problem because solid spheres of different sizes are used to construct models of liquids, of crystals, of colloidal systems, of granular and powder systems. Moreover, the "packaging" of solid spheres is correlated to important problems in information and optimization theory.

IT WAS GALILEO GALILEI who found one of the most powerful tools for investigating nature: simplifying phenomena. He constructed a theory in which friction was completely ignored—and yet in a friction-free world we could not even walk (we would slip and slide) or eat (the food would simply fall off our cutlery). In other words, the Galilean world with which modern physics began is completely different from the real world in which we live. Over centuries we have added other elements so that today we are in possession of a reasonable approximation of the world. This point of view is well summed up in a letter by the sixteenth-century Italian physicist Evangelista Torricelli (a follower of Galileo and author of *De motu* or "Concerning Movement"), in a beautiful passage on the motion of bodies:

> Whether the principles of the doctrine of *De motu* are true or not matters little to me, since, if they are not true, we can pretend that they are as we have supposed, and then take all of the other speculations derived from these principles, not as mixed things but as pure geometry. I pretend or suppose that a given body or point moves downward or upward with the known proportion and horizontally at an equal speed [in modern terms, moving in the absence of atmospheric friction]. When I have said this, everything else that I and Galileo have said follows. If then balls made of lead, of iron, of

stone do not observe these supposed proportions, we can say that we are not talking about them.

For Torricelli, who was also an experienced experimental physicist, it was clear that understanding the motion of bodies in the absence of friction was preliminary to understanding the motion of bodies with friction, and hence that it was a necessary stage. Then, with the passing of decades, it was possible to factor into the equations of motion the friction that Galileo had neglected.

Starting from this reduction of physical phenomena to their essentials, we have developed the physics of recent centuries. And physics has become so potent and rich that it can introduce into its own models complexity and disorder, and study phenomena that Galileo could never have imagined would be part of physics.

Looking back on this fascinating story of the irruption of complexity into physics, I have to say that I have been doubly fortunate. Two Nobel laureates, along with many others, did the hard work of synthesis and formulated the problem in such a way that it was possible for me to make a decisive step forward in just a few months. Furthermore, having flitted between different problems for almost ten years, I had developed the tools that enabled me to understand what was incomprehensible to many—and had acquired a curiosity for problems in areas unfamiliar to me. I had alighted on the right problem at the right time, at the height of my scientific abilities and curiosity.

Six

METAPHORS IN SCIENCE

Given that I was successfully moving from one area of science to another, it came naturally to me to wonder what tools I possessed that allowed me, often intuitively, to develop new ideas. Thomas Kuhn has written that scientific revolutions consist of a change of the "glasses" (that is, the paradigm) through which scientists see the world. Clearly these are complex glasses, compounded of many things, some easily identified, others more concealed. Metaphors, for instance, the importance of which is often overlooked.

Science is founded on experimental evidence, on analytical proofs, on theorems. At the origin of scientific construction, however, there is a great constellation of intuitive reasoning. In science, as in art and many other human activities, intuition comes first, certainties later. We saw this at work in the previous chapter, and I would like to add the following two examples that are emblematic of this process.

When Enrico Fermi and his collaborators discovered that slowed-down neutrons were much more effective than fast neutrons at inducing the radioactive transmutation of many elements, the key to the discovery was the replacement at the beginning of the experiment of a lead brick, which served to shield the neutrons, with a brick made of paraffin. Fermi did this on impulse, without reflection, and as a consequence of this change observed an impressive increase in signal to the radioactivity counters (by more than a hundredfold). His collaborators Edoardo Amaldi, Bruno Pontecorvo, Franco Rasetti, and Emilio Segrè (members of the Fermi group in Rome) were stunned. Fermi quickly explained that the paraffin had slowed down the neutrons and that slow neutrons must be much more effective than fast. And when Amaldi asked, "How did you get the idea of replacing the lead with paraffin?" Fermi replied: "With my remarkable intuition."

Claudio Procesi, my colleague in the Lincean Academy, claims that the difference between a good mathematician and a bad one is that a good mathematician understands immediately which mathematical statements are true and which are false, whereas a bad mathematician has to try to prove them in order to know which are true and which are false.

In both these examples, intuition is extremely important. The tools used go far beyond formal logic, and it is extremely interesting to investigate the intuitive logic that provides the basis of scientific progress and to look at the metaphors, for instance, that play a decisive role in the transfer of images and ideas between disciplines in any given historical period.

If we look carefully enough at such a defined period, it is possible to discern something like a spirit of the age. Frequently we are able to find correspondences and assonances not just between different scientific disciplines—what would become biology, physics, and so on—but also between music, literature, art, and science. It is enough to think of the crisis in a certain kind of rationalism at the beginning of the twentieth century, and of the changes that took place simultaneously in painting, literature, music, physics, and psychology. All these disciplines, which are well defined and separate, nevertheless communicate with each other, and it is reasonable to assume that metaphor plays an important role in that communication.

Unfortunately, in science in general and to an even more exaggerated degree in "hard" science, there is frequently little trace of the intermediate stages required to obtain a result, leaving us unaware of what inspired a scientist for any given idea. Especially in mathematics but also in physics and other scientific disciplines, extrascientific considerations do not make it into formally composed articles. Such texts have been completely purified, reduced to formal language in which there is seldom any allusion to nontechnical reasoning. Sometimes traces of prescientific arguments can be found in texts of a more general nature (in those of Henri Poincaré, for instance), where metascientific reasoning occurs. But in almost all texts written by scientists, these themes are taboo.

. . .

WHILE SEEKING CONCRETE EXAMPLES of how ideas are transferred between different disciplines, I began to reflect on the uses of probability in science. One of the first areas in which probability was used, apart from dice and card games, was statistics—the science of states, as its name suggests—and in the nineteenth century many economists and sociologists, such as Adolphe Quételet, made very important contributions to statistics and to the calculation of probability. In the second half of the nineteenth century, James Clerk Maxwell and Ludwig Boltzmann, apparently independently, created statistical mechanics introducing probability and statistics into physics at the microscopic level, with the express purpose of understanding collective behavior. The Darwinian mechanism of natural selection was also formulated at this time. Genetic characteristics mutate in a random way, and these mutations subsequently come to be selected. For Darwin, the key to the theory of evolution was the concept of selection from among various possibilities.

With the rediscovery of Gregor Mendel's work at the beginning of the twentieth century, the physical substrate on which evolution operates came to be identified with genes; Darwinian theory became the dominant paradigm of biology. It is striking that, at the end of the 1920s, the Copenhagen school's interpretation of quantum mechanics, a field extremely distant from biology, shows a marked similarity to Darwinian selection. A quantum system can be found in various states, and the experiment (or observation) randomly selects one from the different possibilities.

As in Darwinian theory and quantum mechanics, evolution (both biological and physical) passes through an array of new possibilities to a subsequent selection. Obviously the details are fundamentally different: in natural selection the new possibilities arise in a random manner and the choice is deterministic (simplified in the phrase "survival of the fittest"), whereas in quantum mechanics the state evolves in a deterministic manner and the measurement randomly chooses among the various possible outcomes of the experiment. But beyond these differences there is a strong similarity between the two processes, and it is possible that Niels Bohr, Max Born, and the other members of the Copenhagen school had the Darwinian theory of evolution in the back of their minds, and could not fail in some way to be influenced by it. Unfortunately there is no clue to this connection in the known technical works translated into English. Not being a historian of the group, I cannot say for certain that they did not talk about it in some little-known writing. But it is also possible that the scientists themselves did not recognize the weight of Darwin's influence, and never wrote anything about it.

IT IS NECESSARY TO MAKE a very clear distinction between the use of metaphor as a heuristic tool and the use of metaphor, assonance, and other rhetorical figures as the basis for reasoning, to the extreme of even replacing logic with rhetoric. I find this second way of proceeding wholly pernicious; concepts that cannot be translated are transposed into a completely different language, becoming deformed in the process. It is no wonder that conclusions arrived at in this way are often completely arbitrary. Monsters are sometimes created in this translation. In sociobiology, for instance, biological arguments and metaphors are translated, without control, into the social field where they have no application, and the resulting implicit hypotheses are quite incorrect. Dangerous conclusions are reached—and used politically to arrive at such aberrant theories as social Darwinism.

Frequently this casual use of metaphor can also be found in the humanities, with equally negative if less dangerous results. It is difficult at this point not to mention an infamous hoax. In order to satirize pseudo-philosophical-scientific publications of a certain kind, the American physicist Alan D. Sokal concocted an article using the metaphorical style of intellectuals such as Jacques Lacan and Jacques Derrida and their academic followers. The article, entitled "Transgressing the Boundaries: Toward a Transformative Hermeneutics of Quantum Gravity," was based on a series of physical, sociological, and psychological metaphors so nonsensical that if he had used them in earnest his colleagues would surely have become concerned for his mental health. Knowing full well that what he was writing made no sense, Sokal constructed a series of wild comparisons, supported by a formidable critical apparatus and couched in a particular, refined, all-too-recognizable academic style. Almost incredibly, the article was accepted for publication by the editorial board of *Social Text*, one of the leading journals in its field. A scandal ensued when Sokal revealed publicly that he had written and prominently published something that was deliberately nonsensical. Enormous embarrassment ensued, hardly diminished by someone's attempt to argue that the piece probably did make sense, above and beyond the mere intentions of the author. The article, which is available online, makes for an amusing read, and anyone capable of understanding its physical metaphors cannot fail to admire the almost inexhaustible inventiveness of its author.

Despite the misuses highlighted by Sokal, metaphor has an extremely important role in scientific communication, not least when we want to explain a discovery to the public. Frequently, however, metaphors reemerge in common parlance in an unbearably imprecise way. And it is quite natural that metaphors should turn out to be not very faithful: this is what often happens when the words of one language are used in another with a different meaning. This phenomenon, however understandable, tends to make scientists nervous.

I find expressions such as "It's written into our DNA" especially annoying. Every time I hear them, I cannot help thinking that DNA is the basis of the genetic transmission of characteristics, a Darwinian transmission, whereas culture is transmitted in a completely different way, through acquired characteristics—transformations, in other words, that pass in a Lamarckian way from parent to child. Thinking that culture can be transmitted via DNA clashes with the basic principles of the theory of evolution.

IN PHYSICS, on the other hand, metaphor is used most often in situations of crisis, in fierce metascientific discussions, when it is not clear what the physical laws should be. Let's look at some examples.

Einstein was not at all satisfied with quantum mechanics, despite having contributed more than anyone to its birth. For Einstein, quantum mechanics was not the whole story. He contested in particular the Copenhagen school's interpretation, in which probability plays a fundamental role: physical theory *just had to be* deterministic. This is how his famous "God does not play dice" quote came about—to which Bohr apparently replied: "Einstein, stop telling God what to do or not do."

At the end of the 1950s it was discovered that weak interactions (the forces responsible for radioactive decay) do not conserve parity: in other words, in a film of an experiment on weak interactions, we can tell if the film is correct or if right and left have been inverted. This was a completely unexpected result, because the forces of nature do not distinguish between right and left. There was much bewilderment, summed up well by Wolfgang Pauli: "I'm not that surprised if God is left-handed, but I refuse to believe that he is a weak left-hander."

Sometimes it is difficult to understand whether certain arguments work through metaphor or analogy. In the seventeenth and eighteenth centuries, physics was dominated by mechanicism: every physical law could be explained in terms of machines, albeit frequently invisible or microscopic ones. The machines worked through interactions between parts that were in contact with each other. From this conceptual point of view, interaction at a distance was completely inexplicable. The same Newton who proposed the laws of universal gravity (which state that distant bodies are attracted on account of gravity, even though they do not touch—and even, like the planets around the sun, at very great distances indeed) excused himself by saying "Hypotheses non fingo," or "I frame no hypotheses," implicitly leaving it to others to ultimately make sense of the mechanical model.

Gravity as a force acting over distance remained a scandal for more than a century, so much so that numerous attempts were made to explain it within the framework of mechanicism. In one such attempt, perhaps the most ingenious, space was supposed to be filled with an all-pervasive radiation, and objects were pushed by this radiation. Normally the radiation came from all directions and the force exerted was balanced. But when two objects are in the same vicinity, one casts a shadow over the other, and the radiation pushes them closer together. This was believed to be the origin of the force of gravity. The basics of mechanicism survived until the beginning of the twentieth century: the vacuum (ether) became another mechanical factor, with its fluctuations interpreted as the cause of electric and magnetic fields.

IN BIOLOGY TOO we find persistent metaphors that have had an important role to play. In the seventeenth century, for instance, organisms were viewed as machines with very small components—so small they could not be seen. In the second half of the twentieth century, after the discovery of the fundamental role of the information coded in DNA, the computer metaphor was introduced, in which the hardware is the protein system and the software is the DNA. This metaphor was hugely successful due to its explicatory power and the fact that it encapsulated well our knowledge at the time. Later we understood that the interaction between proteins and DNA was much more complex: DNA modifies itself, and subsequent discoveries gradually rendered the computer metaphor obsolete, even though it is still being used.

Some current metaphors in biology are based on complexity, on the idea that from large numbers of interacting agents—molecules, genes, cells, animals, species, depending on the level of discussion—new phenomena are emerging as a result of that collective interaction. There is a tendency to focus on these phenomena and to try to explain them using metaphors and ideas drawn from physics. Among many important ideas, networks stand out (metabolic networks, for instance), as well as the fractal geometry of shapes in nature—from lungs and tree branches to the structure of a cauliflower.

Physics typically makes great use of models, and models are themselves a type of metaphor. I was struck by a discussion between two friends of mine about physicists' resistance to metaphors and our tendency to dismantle them. In short, one argued that the comparison between the motion of wheat swaying in the wind and the waves of the sea is not metaphorical, inasmuch as the equations that describe those marine waves are similar to the equations that describe the movement of the wheat. In the final analysis, we are dealing with the same phenomenon, not with a metaphorical comparison. The other friend noted that, for the overwhelming majority of people, the undulation of the wheat and the waves of the sea seem like two inherently different phenomena.

What is behind the tendency of physicists to distrust and dismantle metaphors? To answer this question, we need to reflect on what physics is as a science, and how it compares with mathematics and the other natural sciences. Physics can be considered a kind of applied math. It starts with a concrete problem and translates it into the language of physics, which, ever since Galileo, is mathematics. The physicist sometimes uses mathematics ungrammatically; not following all the rules of grammar is a license that we grant to poets.

But then what exactly is mathematics? It's a science that operates through symbols that have been purified of any concrete meaning, or as Bertrand Russell put it, "Mathematics may be defined as the subject in which we never know what we are talking about." The reason for this is simple: if I state that 2 + 3 makes 5, it could be 2 telephones + 3 telephones that makes 5 telephones, or 2 cows + 3 cows that makes 5 cows. We have no idea what the objects in question are. This is true at an extremely low level of abstraction, and becomes more and more relevant as we move toward more abstract concepts. Mathematical objects are purified of all physical form, and hence mathematical propositions, like logical ones, have a universal value.

A physicist, on the other hand, translates concrete phenomena into a mathematical language where many of their material characteristics are lost, leaving just those essential for studying the phenomena. The undulation of wheat and ocean waves comes to be described by very similar equations, and after they have been represented with the same equation, one is no longer a metaphor for the other. They become different physical incarnations of the same mathematical representation. In reality the equations are not exactly the same but belong to the same family, which is to say they both describe the propagation of waves. In the case of the wheat, the speed of the propagation of the waves is independent of their length—the distance between two successive waves—whereas in the case of marine waves, the velocity is approximately proportionate to the square root of the length of the wave. This is why the waves of a tsunami, which are extremely long, travel so extremely fast.

THAT COMPLETELY DIFFERENT SYSTEMS could have the same mathematical description was a very important discovery for physicists. Sometimes, however, the equations are the same but the mathematical expressions that correspond to the observable quantities are different. In these cases—the most interesting—the observed behavior of two systems can be very different, and they may also belong to separate fields of physics (solid-state physics and particle physics, for instance), so their coming together in the same mathematical representation can be completely unexpected.

From the moment we realize that two systems in very different fields of physics can be described by the same mathematical structure, a rapid advancement of knowledge takes place in which the two fields cross-fertilize. When the two systems have been well studied, it is possible to apply to one field (after an appropriate translation) the myriad results and techniques obtained in the other field. In general, when the same formal mathematical system has two completely different physical manifestations, you can use physics insights from both systems to obtain valuable complementary information.

In a work written in 1961 with Yoichiro Nambu, Giovanni Jona-Lasinio described an analogy between the quantum void and superconductivity. The use of the word "analogy" is actually very dated. Between the mid-1960s and the 1970s we realized that the calculation of the statistical properties of a material and the structure of the quantum void are two different aspects of the same mathematical problem. The information that came from experiments performed on metals (that certain materials were superconductors, for instance) shed light on the possible behavior of the quantum void. From the 1980s onward, the word "analogy" had disappeared, replaced by phrases such as "We can conjecture that the quantum void is superconducting."

The relation between the statistical mechanics of materials and the quantum physics of elementary particles was really important. Perhaps the most striking example of this relation is the work begun by Jona-Lasinio and Carlo Di Castro that first applied renormalization group (RG) to the study of phase transitions. In fact, as we have seen, renormalization group was developed in the context of quantum and relativistic field theory, and all the techniques refined in that context have been applied to the statistical mechanics of critical phenomena—with great success, as evidenced by the Nobel Prize being awarded to Kenneth Wilson. The techniques based on RG were key to understanding critical phenomena and were subsequently reported in elementary particle physics. During this to-and-fro they were enriched with new ideas and with a new physical understanding of phenomena, and it was from this moment onward that RG began to have a fundamental role in the physics of elementary particles.

It does not seem to me that we can speak of metaphor in such cases, for this cross-fertilization is very different from the traditional rhetorical figure. The same mathematical abstraction can be projected onto disparate physical systems, and each of these perspectives enlightens us on different aspects. Take complex systems, for example, meaning systems made up of many agents. Sometimes the same mathematical model can be applied to the study of the behavior of exotic magnetic systems at low temperature (spin glasses), to the functioning of the brain, to the behavior of large groups of mammals, and to the economy. In a case such as this, using conclusions that come from one field in order to make predictions in another is not exactly having recourse to a metaphor, because we are dealing with systems that have a similar mathematical formalization. It is, rather, an attempt to carry concepts from one discipline to another, an attempt that is justified by shared correspondence with the same mathematical structure.

Seven

HOW IDEAS ARE BORN

In 2007 I was lucky enough to be invited to Bari, in southern Italy, for a public debate on creativity with Gianrico Carofiglio, the anti-Mafia judge turned bestselling writer of crime fiction. It took place in a room packed with young people. The theme was "Where do ideas come from? A conversation between a physicist and a writer." Even though we had such different backgrounds, in the course of our discussion we discovered much that we could agree about. We both emphasized the importance of talking to others. We discussed the danger that our thinking could become too fixed to the point of hindering new ideas. This is as true in physics as it is in the interpretation of the world, which, according to Carofiglio, is the subject of writing. The writer must constantly stay open to the outside world. Perhaps it is easier for the scientist, especially if they are in touch with new generations of researchers who are not too rigid and can come up with unexpected innovations. In the end, both of us believe good communication involves formulating our ideas clearly and precisely, which often cannot be done without revisiting what we took for granted.

For me it was an important occasion that set me thinking about some very interesting questions that I might be able to help answer. How are ideas formed in the mind of a theoretical physicist like me? What types of logical procedures do we use? I am not just talking about those grand ideas that can modify the history of thought and humanity. I want to talk instead about what has been called "microcreativity," those small everyday ideas that are crucial to making any progress in a scientific context. For me an idea is an unexpected thought—one that is surprising and by no means banal.

The birth of ideas is a subject that many scientists have confronted in their writings. Let's begin with the famous French mathematicians Henri Poincaré and Jacques Hadamard. The two men, who lived in the nineteenth and twentieth centuries, repeatedly described the ways in which their mathematical ideas were born, and had a common point of view on the subject: there are distinct phases to the formulation of a theorem.

- There is a first, preparatory stage in which the problem is studied, the existing literature read, and the first unsuccessful attempts at a solution are made. It is a period that can last between a week and a month, and ends because no progress occurs.

- Then there is a period of incubation in which the problem is abandoned, at least consciously.
- This incubation ends suddenly with a moment of illumination, which often occurs in a situation unrelated to the problem you're trying to solve. It might happen, for instance, when conversing with a friend about topics with no apparent connection to the problem.
- In the end, after the illumination provides the general way to tackle the problem, the solution must actually be formulated. This can be a very protracted period. You must verify that your idea is correct, and whether or not the road you have set out on can work—followed by all the mathematical steps required to demonstrate the solution.

Obviously sometimes the moment of illumination turns out to be misleading, by assuming the validity of steps that cannot in fact be taken. In these cases you have to start all over again.

It is a very interesting description of the process, and one that assigns a prominent role to unconscious thinking. Even Einstein was in agreement with this. On various occasions, he underlined how important unconscious thinking was to him. Pausing one's concentration on a difficult problem—to allow ideas to *settle*, and to face that problem again with a fresh mind—is no doubt very common. The Italian phrase *La notte porta consiglio* has counterparts in so many languages: *In nocte consilium* ("Night is the time for counsel"—or, more colloquially, "Sleep on it"); *Die nacht bringt rat* ("Night brings advice"); *Il est utile de consulter l'oreiller* and *Antes de hacer nada, consúltado con la almohala* (where *l'oreiller* and *almohada* both mean "pillow," so literally "It is useful to consult your pillow"); *La note xe la mare d'i pensieri* ("Night is the sea of thought").

Passing from big to more mundane problems, I'd like to add an experience of my own. Very often my research in theoretical physics requires writing computer programs, a task I find both enjoyable and relaxing. The computer is a machine completely lacking in common sense, and therefore it will do exactly what you tell it to do, adhering to your instructions with maddening precision, to the letter. If you tell a human being to take a particular road and then continue straight on, with any luck they will not leave that road at the first curve. For the computer, on the other hand, such behavior would be perfectly normal unless you have spelled out very precisely what you mean by continuing *straight on.*

What you first ask a computer to do is often subtly different from what you really intended to ask. A new program, written in one of the many programming languages, frequently does not work: if we do simple tests, the results are completely different than expected. Or at any rate this is my experience. Obviously the better you are at programming, the more likely it is that your instructions will hit the mark the first time.

On countless occasions I have struggled all morning to discover the mistake I made. I read the program carefully and think about all the instructions one by one, asking myself whether a comma is in the right place, if a semicolon is missing, if there is one too many or too few equal signs—without figuring it out. Then, as I'm halfway home in the car, I suddenly see that it's *there* where I have gone wrong, and on arrival I hurry to verify that the correct solution has come to me while driving.

This is a very common experience. Another time—perhaps the only time in my life—a similar but much more spectacular episode of this kind occurred. Together with colleagues, I had tried, without success, to come up with a strategy to solve a very difficult problem. For a long time (ten to fifteen years), various approximate solutions had been proposed. I had worked on the problem myself, but then abandoned it because it seemed too difficult. Until one day, over lunch at a conference, a friend said to me: "You know, the problem that you've been working on is very interesting, because its solution would have a number of applications above and beyond those previously anticipated." To which I replied: "In that case more effort should really be made to solve it. Maybe if one were to begin by . . ." And I then explained to him, step by step, a fully formed strategy for solving the problem—a strategy that turned out to be both complete and correct.

IT IS EASY TO recognize in these examples the process of incubation. I'm convinced that we all have similar anecdotes to tell. But if incubation, whether for small matters or large, is an unconscious process, we then have to ask ourselves what kind of logic it follows and how it comes about. Very often we take for granted that thought is verbal and that unconscious reasoning is not. Einstein would not have agreed, arguing that being fully conscious is one end of a spectrum that is never actually reached: there is always, in all thinking, an admixture of the unconscious.

Even though I am not an expert in the field, allow me to offer some of my own observations on conscious and unconscious thought. We have the impression of thinking using words, formulating sentences. This is the case not just when we talk to others, but also when we reflect in silence. If someone asked us to reflect on a problem without using words, we would feel completely helpless: we are not capable of solving problems in our heads without formalizing or embodying our reasoning in words; they can be words in any language, but there have to be words.

The way in which we think, however, is not completely based on words. In fact, when we begin to think or say a sentence, we need to know where we are headed. There are grammatical rules we must follow. We don't begin a sentence with the word "not" and then stop, not knowing what to say, because the moment the word comes to mind, we already know the next word and probably the whole sentence as well. But if so, the whole sentence must be present in our mind in nonverbal form before it is expressed in words.

Formulating thoughts through words is extremely important; words are powerful—they link together and attract one another. They have the same function, fundamentally, as algorithms in mathematics. Just as the algorithm carries out mathematical reasoning almost by itself, so words have a life of their own; they evoke other words, and they allow us to make abstractions and deductions and to use formal logic. Perhaps this process of attaching language to our thoughts is also useful for memory: if we do not formalize our thoughts in words, they are more difficult to remember. Verbal thought, however, must be preceded by nonverbal thought. This statement is not so strange considering that thought is much older than language. Human language is probably tens of thousands of years old, but it is difficult to believe that before language our ancestors did not think—or that animals, or human infants who do not talk yet, do not carry out some form of thinking.

Unfortunately it is quite difficult to understand what type of logic nonverbal thinking follows, not least because logic refers to language, and it is almost impossible to study a nonverbal thought using the tools of language. Unconscious thinking, however, is crucial to formulating new ideas: it is not only used during the incubation period that Poincaré and Hadamard talked about, but it is also the basis of the more general phenomenon of mathematical intuition. In fact, at first glance, mathematical intuition exhibits some surprising characteristics.

The proof of a theorem often consists of many successive steps, at the end of which we arrive at a solution, deduction after deduction. Except in rare cases, however, this is not how the theorem is demonstrated in the first place. Usually the statement is formulated first, then, knowing where to begin and where we will end up, the intermediate steps are established and connected to each other with the necessary demonstrations, until we arrive at the complete proof. And in this way a bridge is built: at first we decide where we want the construction to start and end, then we lay the foundations of the intermediate pilings, and finally we establish the road. It makes no sense to build the first section of a bridge before even thinking about planning the second, risking the discovery only then that it is not possible to lay the foundations for it.

Just as a sentence must be present in its entirety before it is formulated in words, a demonstration must be present in the mind of a mathematician, at least in outline, before passing to the deductive stage.

This way of proceeding explains why there are so many valid theorems in which the first proof was wrong. Often a mathematician, after correctly formulating a theorem and identifying a possible path, gets an intermediate stage wrong. If the intuition was more or less right, either there is another, correct way of following the difficult stage or there is a different way of arriving at the same result. Mathematicians often speak of the "meaning" of a theorem—meaning that is articulated in an informal language based mostly on analogies, similarities, metaphors, intuitions. But there is often no trace of this meaning in mathematical texts, which use a different language: the meaning justifies in some way the original intuition, but not being formalizable, it is felt to be imprecise—something you might talk about with friends but would not think of incorporating into a text that is required to be rigorous.

THERE IS ALSO INTUITION in physics. It is different from mathematical intuition and has evolved over time. Galileo, as the historian of science Paolo Rossi has pointed out, had the grand intuition that the celestial and the terrestrial worlds were similar, and that it was possible to apply the same laws to both. This was the starting point for many of Galileo's discoveries, though it is not easy to prove, since reasoning often has its tail in its mouth, so to speak, as the irreverent philosopher of science Paul Feyerabend has emphasized. Spots on the sun demonstrated that the celestial world was corruptible, but only if they were not an effect produced by the telescope itself. Given that it was not possible to establish whether the telescope accurately depicted the heavens, Galileo's observations implied either that sunspots existed and that the celestial world was mutable like the terrestrial one, or that the telescope produced false images and interacted differently with light from terrestrial and celestial objects. This second hypothesis is evidently very difficult to sustain inasmuch as the sunspots rotate at a constant speed (due to the rotation of the sun). At the time, however, the idea of one set of rules for the whole universe was shocking, and many were unable to accept the Galilean intuition, rejecting what followed from it.

Intuition in physics also had a fundamental role to play much later, particularly during the birth of quantum mechanics at the beginning of the twentieth century. This was one of the great adventures of physics, and between 1900 and 1925 involved such illustrious scientists as Max Planck, Albert Einstein, Niels Bohr, Werner Heisenberg, Paul Dirac, Wolfgang Pauli, and Enrico Fermi. They were dealing with an apparently very strange process, in some respects a contradictory one. Certain phenomena had been observed (blackbody radiation, for example) that physicists at the time were unable to understand because the phenomena could be explained *only* by the quantum mechanics that had not yet been discovered.

What would be the most logical way of proceeding? To invent quantum mechanics and present the right explanation! But the story took a completely different turn. Physicists made various attempts to explain quantum phenomena in classical models by explicitly assuming that some of the lesser known elements of the model behaved in a bizarre way (incompatibly with classical mechanics)—I believe that they were thinking, "There are things we do not yet understand, but we will after more work." During this time there were a large number of contradictory contributions to the field, some of which were plainly wrong—though to be fair they could hardly be right because, by trying to justify quantum phenomena within the framework of classical mechanics, they were attempting the impossible. For example, in a 1900 article that sought to explain blackbody radiation, Planck assumed that light interacted with oscillators that had the correct quantum properties, in complete incompatibility with classical physics. Planck plowed ahead, having not realized that the supposed compatibility with classical physics did not in fact exist.

It is remarkable to see how the partial explanations presented were in fact correct: intuition in physics is so powerful that hypotheses that remained in the field of classical mechanics contributed to explaining quantum phenomena, pushing ever forward the contradictions between classical mechanics and observed phenomena. In the end, when the contradictions were too great, many aspects of the new quantum mechanics had already been anticipated. To give one example, in Bohr's theory of 1913, which assumed that the electron that orbits the hydrogen atom could do so only on *certain* orbits that satisfy a certain condition, the spectral lines of the light emitted by hydrogen could be calculated in a simple way. The hypothesis was not sustainable within classical mechanics, but it provided fundamental clues that helped build quantum mechanics a decade later, when the urgent need for such a radical new framework became clear.

The final barriers fell in 1924 and 1925; the following years saw progress at an impressive rate, and by the end of 1927, the new quantum mechanics had virtually reached its definitive formulation. The preparatory work (which lasted twenty-five years, from 1900 to 1925) was possible precisely because there had been a strong intuition as to how the physical system was organized. It was a very different kind of intuition from that of the mathematician's, leading to work that advanced physics despite being frequently based on wrong arguments.

Concerning intuition, a friend of mine who is an experimental low-temperature physicist once remarked: "You have to get to know your experimental setup so well, the system that you are measuring, the phenomena that you are observing, to be in a position to give the right answer without even thinking. If they ask you a question (or you ask it), you must be in a position to give the right answer immediately and then afterward, on reflection, to be able to say why it is right."

I recently experienced something that points in the same direction. A friend I work with asked me a not very easy question, to which I immediately gave a detailed answer. Then he asked me how I'd arrived at it. At first I gave a completely nonsensical explanation, then a second that made a bit more sense, and only at the third attempt was I able to properly justify the right answer, which I had at first given for the wrong reasons. Italian physicist Giovanni Gallavotti, in the preface to his book on mechanics, writes that good students must reflect on a theorem until the theorem seems obvious and the proof, consequently, seems superfluous.

Intuition depends very much on the field in question; in some cases it is based on mathematical formalism. Formalism is an extremely powerful tool, and becomes even more so if the unconscious itself gets accustomed to using algorithmic procedures. As we have seen, when I was doing my first research on spin glasses, I used the replica method, a pseudo-mathematical formalism (in the sense that the mathematical validity of the method had not been proven but the results were correct, as was demonstrated twenty years later) that allowed me to arrive at a final result without knowing what I was doing. It then took years to understand the physics significance of my results. I had unconsciously constructed a series of rules that allowed me to understand the direction in which to proceed with the calculations. Rules that I would never have known how to formalize.

Making progress without being fully conscious of what you are doing is hardly a method confined to scientific problems. The great twentieth-century Italian writer Luce d'Eramo, whose books have been translated into many languages, said that when writing a novel she often proceeded by rereading everything she had written so far, and only then did she decide how to begin the next scene. At that point she took the characters, mentally inserted them into the scene, and observed them: "I don't decide what they have to do, I imagine them and observe how they speak and how they act. I merely record what they do." It's a process that is not a million miles from the one described by Poincaré and Hadamard.

I WOULD LIKE TO give one last indication as to how our reasoning is more complex than we think. I have always been struck by the difficulty of demonstrating the truth or falsity of a proposition when we have no clue as to the final results that will be yielded by testing it. If there are strong heuristic arguments that imply that a statement is true (or false), then often (but not always) it is *easier* to find the proof. On the other hand, if we lack indications either way, we can expect to take twice as long to reach a final result: half the time reasoning as if the result is true, and half the time assuming it is false. This is easily said, harder to actually do; in practice we seek arguments to demonstrate the truth of the proposal, and if we do not succeed we try to prove its falsity, oscillating between the two without getting much further forward. Perhaps we can consciously move from one hypothesis to its opposite, but the unconscious remains confused.

The effect of a little additional information can be underlined by an episode that I witnessed and that left me stunned. A very interesting property (let's call it X for simplicity's sake) had been verified in the context of extremely simplified models, and it was crucial for the development of the theory to understand whether the property could be proved for more realistic systems: its validity allows experiments on materials to be carried out to verify detailed predictions of the theory. My friends and I had been talking about it for years: no one had any idea how the demonstration might be achieved, and we doubted that the property was demonstrable, or even if it was real.

One day my friend Silvio Franz told me that, together with Luca Peliti (two brilliant physicists with whom I have collaborated in the past), he had proved the property X by using a very simple but extremely astute idea. I was happy with the proof; I went to Paris and during a conference declared that I had complete confidence that the property X was demonstrable. I didn't announce the result because I wanted to wait for my friend to write it up. After the conference, another friend, Marc Mézard, said to me while we were on the steps of the École Normale: "Sorry, Giorgio, but why have you said that the property X can be proved?" I replied that it had already been proved by Franz and Peliti. To my great surprise, Mézard instantly said, "Oh, yes, I've seen too," and briefly, then and there, he summarized the correct demonstration. The simple information that the property was demonstrable was enough for him to arrive at the long-sought-after proof for himself in less than ten seconds.

It is striking how sometimes a minimal amount of information is enough to cause substantial progress in a field to which much thought has been given. For example, in 1907, Einstein had been thinking a lot about gravity and one day had "the happiest thought" of his life: when we plunge into free-fall, we no longer feel the force of gravity, which is nullified within us. The force of gravity, in other words, depends on a system of reference, and by choosing an appropriate system of reference it is possible to cancel it out, at least locally. Starting from this observation, he was able to build the theory of general relativity, which is perhaps his greatest contribution to physics and the one that was most ahead of its time.

It's said that Einstein had this intuition following a curious incident. Whether true or not, it is certainly fitting. A house-painter was decorating the outside of Einstein's apartment building. Sitting on a chair on scaffolding while working on the third floor, the painter somehow lost his balance and plummeted to the ground while still seated on the chair, fortunately escaping with just a few broken bones. Several days later, Einstein remarked to a neighbor: "Who knows what that poor housepainter must have been thinking as he fell," and his neighbor replied: "I spoke to him, and he said that as he fell he did not feel as if he was still on the chair, almost as if there was no force of gravity." This sparked something for Einstein, and starting from the painter's observation, he was able to arrive at his celebrated theory. It is remarkable how the origins of the theory of gravity have been connected with things falling: an apple for Newton, a painter and decorator for Einstein. In both cases, however, it would take many years of hard work to convert felicitous insight into physics theory. There are many who have had the right intuition but have then been unable to bring it to fruition.

Eight

THE MEANING OF SCIENCE

"Science is like sex: sometimes something useful comes out, but that is not the reason we are doing it." So said Richard Feynman (allegedly), one of the greatest physicists of the twentieth century and perhaps the most charismatic.

This sentence, together with the Dantean imperative that we "were not made to live like brutes, but to seek virtue and knowledge," sums up pretty well the motivating, subjective passions of scientists. Science is an enormous puzzle, and every piece of it that is put correctly into place opens the possibility of connecting other pieces. In this gigantic mosaic, every scientist adds certain tesserae, with the knowledge of having made such a contribution and of the process whereby, when our own names are forgotten, those who come after will nevertheless be standing on our shoulders in order to see further.

We can imagine a vivid metaphor for the scientific enterprise: Some sailors land on an unknown island at night and light a fire on the beach. They begin to see their surroundings, and the more wood they add to the fire, the greater the area they can see. But beyond that area there remains a mysterious region that can hardly be made out in the almost-complete darkness, barely broken by the light of the fire. That area becomes dimmer with distance, and yet it expands as the bonfire builds. The more we explore the universe, the more new regions we discover to explore: every new discovery enables us to formulate so many new questions that up until then we were completely unable to ask.

Beyond such considerations, however, it is fundamental that scientists find seeking to solve the puzzle *fun*. When I was discussing potential assignments with my teacher Nicola Cabibbo, he would ask: "Why study this problem if it doesn't really grab you?" Often among scientists, there is a feeling verging on astonishment that we are being paid to do what we are so passionate about. Or as my dear friend Aurelio Grillo used to say: "Being a physicist is hard, but it's better than having to work for a living."

Nevertheless, except in those rare cases in which the scientist is from a wealthy family and their research is conducted during long periods of leisure (Pliny the Elder or Fermat, for example), the scientist has indeed always had the problem of earning a living, and the applications of science have been crucial not least for this reason. Think of one of the first sciences in history: astronomy. It is hard for us to imagine, living in our well-lit cities, the enormous privilege and power of those who had knowledge of the flow of the seasons and the movements of the stars, who could predict eclipses of the moon, not to mention such a tremendous phenomenon as a solar eclipse.

Even if scientists' early motivations had mainly to do with cultural or social prestige, the practical applications of science certainly did not escape them. Galileo, for example, proposed that the eclipses of Jupiter's satellites could be used to determine the exact time without any need for precision timepieces. Knowing the exact time was essential in order to deduce longitude from the sun and other stars, enabling mariners to calculate their position. In reality, Galileo's proposal was too cumbersome to be put into practice, and the problem was definitively solved instead by the introduction of the precision chronometer, crowning more than a century's worth of research.

To coordinate scientific research, numerous academies were established in the seventeenth and eighteenth centuries and still dominate today: the Lincean Academy in 1603, the Royal Society in 1660, the French Academy of Sciences in 1666, the American Philosophical Society in 1743. The latter is particularly noteworthy, having been founded by Benjamin Franklin with the explicit purpose of promoting *useful knowledge*.

With the passage of time, science became ever more useful to society (economic development is based on scientific progress), but also increasingly expensive and demanding of complex equipment and organization. The Second World War marked the first stirrings of so-called big science: Vannevar Bush coordinated the war efforts of six thousand American scientists, while fifty thousand people worked on the construction of the first atomic bombs. Today the research and development sector absorbs just over 1 percent of GNP in Italy, compared with more than 4 percent in South Korea—making it the nation that not only knocked my own out of the World Cup in 2002, but that is now spending three times as much as Italy on scientific research and development.

Science and its institutions need to be funded by society, and society hardly gives a damn about whether the scientists are enjoying their work or not. This point of view was given a textbook airing by the Soviet delegation at the Second International Congress of the History of Science, held in London in 1931. An unexpectedly large Soviet delegation participated, presenting a point of view that was quite different from that of Western scholars. A large part of the delegation disappeared a few years later during Stalin's purges. Nikolai Bukharin, a top-level political player who was very popular in the USSR and subsequently one of the most prominent victims of the purges, spoke at the conference, and wrote in the paper he presented, later published in the book *Science at the Cross Roads*, that "the idea that science is an end in itself is naive: it confuses the *subjective passions* of the professional scientist, who works within a system with a very strict division of labor . . . , with the objective *social role* of this kind of activity—an activity of great *practical* importance."

The development of technology, however, is unthinkable without a parallel advance in pure science. As was clearly shown in

L'ape e l'architetto (*The Bee and the Architect*, published in Italy in 1976 by the four physicists Giovanni Ciccotti, Marcello Cini, Michelangelo de Maria, and Giovanni Jona-Lasinio, and published in English in 2023), pure science not only provides applied science with the knowledge it needs to develop—languages, metaphors, conceptual frameworks—but also has a more hidden if no less important role to play. In fact, basic scientific activities work like a gigantic proving ground for technological products, and for stimulating the consumption of high-tech goods.

This deep integration between science and technology might lead us to think that science has a bright future in a society that is ever more dependent on advanced technology. Today's cell phones, for instance, have the ability to perform hundreds of billions of calculations per second—more or less what it would have taken a mammoth, room-filling supercomputer to achieve in the nineties.

In reality things are very different. There is a strong anti-scientific tendency currently at work; the prestige of science and popular trust in it is being rapidly undermined, and astrological, homeopathic, and aggressively anti-scientific practices (witness the recent anti-vax movement, or the denial that *Xylella fastidiosa* is the origin of the disease afflicting olive trees in Italy, to say nothing of certain approaches to COVID) are spreading fast alongside voracious technological consumerism.

It is not easy to fully understand this phenomenon. Perhaps popular distrust of science is partly the result of a certain perceived arrogance in scientists who present science as a kind of absolute knowledge compared with other, more questionable types of information, even when it is often anything but. Sometimes this arrogance is rooted in scientists who don't make evidence available to the public, instead merely asking for unconditional assent based on trust in experts. This refusal to accept and acknowledge their own limitations can undermine the prestige of scientists who display an excessive, disingenuous confidence to a public that has a perception of the partiality and limits of their views. Sometimes bad communicators present the results of science almost as if it were a superior kind of sorcery, comprehensible only to initiates. In the face of a science perceived as inaccessible magic, nonscientists are pushed toward irrational beliefs. If science comes across as pseudo-magic, then why not opt for actual magic instead?

Technological development needs science, but this need is not enough to guarantee the progress of science. We have seen this in the past. The Romans kept Greek technology without caring much about science, and Christian fanatics commanded by Saint Cyril of Alexandria, holy bishop and Father of the Church, calmly tore apart the mathematician-astronomer Hypatia without worrying about the long-term consequences—rejoicing, rather, in the disappearance of a profane knowledge that was considered useless at best.

Science needs to be defended not just for its practical aspects but for its cultural value. We should have the courage to take as our example the experimental physicist Robert Wilson, who in 1969 when repeatedly asked by an American senator about the applications of the Fermilab accelerator located outside Chicago, and in particular if it was militarily useful in defense of the country, replied, "Its value lies in the love of culture: it is like painting, sculpture, poetry, and like all of those activities of which Americans are patriotically proud, it does not serve to defend our country but makes our country worth defending." This quote from Wilson's congressional testimony became quite popular for its stressing of the cultural values of science.

For science to affirm itself as culture, we must make the public aware of what science is and how science and culture are intertwined, both in their historical development and in the practice of our time. We must explain what scientists do, and what challenges we face, in a way that is not magical or obscurantist. It is no easy task, especially with the hard sciences, where mathematics plays an essential role. The right kind of effort can lead to excellent results.

It is often said that the hard sciences cannot be understood by anyone who has not studied mathematics. But something similar could be said for Chinese poetry, which is an inextricable mixture of literature and painting: the original manuscripts of that poetry are paintings in which the individual ideograms are represented differently each time. These pictorial elements are completely lost in translation, and their beauty cannot be fully appreciated by anyone who does not know Chinese. And yet just as we can get a sense of the beauty of Chinese poetry from a good translation, whether English, Italian, or any other language, so it is possible to appreciate the beauty of the hard sciences even without having formally studied mathematics and science.

It is not easy, but it is possible. We need to promote initiatives that allow people to approach modern science. If we don't do this, scientists themselves will not be able to escape responsibility for the consequences.

I have tried to shoulder my responsibility: this book is my attempt to convey to a wide readership something of the beauty, importance, and cultural value of modern science. If you have followed me this far, I hope that I have succeeded in my attempt.

NOTES ON SOURCES

Before developing it in this form, I had been working for years on this book, which grew out of a series of interviews I did with Anna Parisi. The interviews became sketches for chapters, and I have chosen to collect and develop here only the topics linked to the work for which I received the Nobel Prize in October 2021. Though we share a surname, Anna is not a relative of mine; I have gladly been involved in various projects with her to communicate science, and she helped me draft sections of the book.

Three chapters revisit, with some revisions, pieces that have already been published. "Metaphors in Science" and "How Ideas Are Born" were originally pieces delivered at two conferences of the Lincean Academy in Rome, titled "Metaphors and Symbols in Science" (May 2013) and "The Natural History of Creativity" (June 2009) respectively. These papers were subsequently collected in two volumes published by Scienze e Lettere in 2014 and 2010. "The Meaning of Science" originated as an article with the title "What is science for?" in *Le Scienze* in September 2018, on the occasion of the fiftieth anniversary of the magazine.

Below are the bibliographical details of sources cited in each chapter.

CHAPTER 1: IN A FLIGHT OF STARLINGS

The article in which we published the first results of our research is Michele Ballerini et al., "Interaction ruling animal collective behavior depends on topological rather than metric distance: Evidence from a field study," *Proceedings of the National Academy of Sciences* 105, no. 4 (2008): 1232–37, doi.org/10.1073/pnas.0711437105.

The Max Planck quote comes from M. Planck, *Scientific Autobiography and Other Papers*, trans. F. Gaynor (New York: Philosophical Library, 1949), pp. 33–34.

CHAPTER 2: PHYSICS IN ROME, AROUND FIFTY YEARS AGO

The articles in which Gell-Mann and Zweig independently announced the quark model are M. Gell-Mann, "A schematic model of baryons and mesons," *Physics Letters* 8, no. 3 (1964): 214–15, and G. Zweig, "An SU_3 model for strong interaction symmetry and its breaking," in *Developments in the Quark Theory of Hadrons*, eds. D. B. Lichtenberg and S. P. Rosen, 22–101 (Nonantum, MA: Hadronic Press, 1980), dx.doi.org/10.17181/CERN-TH-412. Color was introduced in O. W. Greenberg, "Spin and unitary-spin independence in a paraquark model of baryons and mesons," *Physical Review Letters* 13, no. 20 (1964): 598–602, doi.org/10.1103/PhysRevLett.13.598.

The metaphor of the pheasant and the veal appears in M. Gell-Mann, "The symmetry group of vector and axial vector currents," *Physics Physique Fizika* 1, no. 1 (1964): 63–75, doi.org/10.1103/PhysicsPhysiqueFizika.1.63.

CHAPTER 4: PHASE TRANSITIONS, OR COLLECTIVE PHENOMENA

Concerning renormalization group, the articles by Kenneth Wilson that I have referenced are K. G. Wilson, "Renormalization group and critical phenomena. I. Renormalization group and the Kadanoff scaling picture," *Physical Review B* 4, no. 9 (1971): 3174–83, doi.org/10.1103/PhysRevB.4.3174; "II. Phase-space cell analysis of critical behavior," *Physical Review B* 4, no. 9 (1971): 3184–205, doi.org/10.1103/PhysRevB.4.3184; "Renormalization group and strong interactions," *Physical Review D* 3, no. 8 (1971): 1818–46, doi.org/10.1103/PhysRevD.3.1818; "Feynman-graph expansion for critical exponents," *Physical Review Letters* 28, no. 9 (1972): 548–51, doi.org/10.1103/PhysRevLett.28.548; K. G. Wilson and M. E. Fisher, "Critical exponents in 3.99 dimensions," *Physical Review Letters* 28, no. 4 (1972): 240–43, doi.org/10.1103/PhysRevLett.28.240.

CHAPTER 5: SPIN GLASSES: THE INTRODUCTION OF DISORDER

The first models of spin glasses were put forward in S. F. Edwards and P. W. Anderson, "Theory of spin glasses," *Journal of Physics F: Metal Physics* 5, no. 5 (1975): 965–74, doi.org/10.1088/0305-4608/5/5/017, and D. Sherrington and S. Kirkpatrick, "Solvable model of a spin-glass," *Physical Review Letters* 35, no. 26 (1975): 1792–96, doi.org/10.1103/PhysRevLett.35.1792.

This is the sequence of my own contributions: G. Parisi, "Toward a mean field theory for spin glasses," *Physics Letters A* 73, no. 3 (1979): 203–5, doi.org/10.1016/0375-9601(79)90708-4; "Infinite number of order parameters for spin-glasses," *Physical Review Letters* 43, no. 23 (1979): 1754–56, doi.org/10.1103/PhysRevLett.43.1754; Marc Mézard et al., "Nature of the spin-glass phase," *Physical Review Letters* 52, no. 13 (1984): 1156–59, doi .org/10.1103/PhysRevLett.52.1156. The book in question is M. Mézard, G. Parisi, and M. Virasoro, *Spin Glass Theory and Beyond: An Introduction to the Replica Method and Its Applications* (Singapore: World Scientific, 1987).

Further applications: G. Parisi and F. Zamponi, "Mean-field theory of hard sphere glasses and jamming," *Reviews of Modern Physics* 82, no. 1 (2010): 789–845, doi.org/10.1103/RevModPhys.82.789.

CHAPTER 6: METAPHORS IN SCIENCE

The article by A. D. Sokal, "Transgressing the boundaries: Toward a transformative hermeneutics of quantum gravity," *Social Text* 46/47 (1996): 217–52, can be read at doi.org/10.2307/466856.

ACKNOWLEDGMENTS

I would not be the scientist I am without the contributions of my teachers, my students, the colleagues with whom I have studied and worked. It should go without saying that research is a collective phenomenon, a complex system in its own right. I have mentioned only some of the people with whom I have worked; to them and to the hundreds of others I might have named, even then at the risk of forgetting someone, I would like to express my gratitude.

INDEX

Page numbers in *italics* indicate figures or photographs.